CHEMISTRY
A Basic Approach

B. RICHARD SIEBRING
Professor of Chemistry
University of Wisconsin

MARY ELLEN SCHAFF
Lecturer
Department of Chemistry
University of Wisconsin

McGRAW-HILL BOOK COMPANY New York St. Louis San Francisco
Düsseldorf   Johannesburg   Kuala Lumpur   London   Mexico   Montreal
New Delhi   Panama   Rio de Janeiro   Singapore   Sydney   Toronto

This book was set in Optima by York Graphic Services, Inc., printed by Halliday Lithograph Corporation, and bound by The Book Press, Inc. The designer was J. E. O'Connor; the drawings were done by Vantage Art, Inc. The editors were James L. Smith, Nancy L. Marcus, and Janet Wagner. Les Kaplan supervised production.

CHEMISTRY: A Basic Approach

Copyright © 1972 by McGraw-Hill, Inc. All rights reserved. Printed in the United States of America. No part of this publication may be reproduced, stored in a retrieval system, or transmitted, in any form or by any means, electronic, mechanical, photocopying, recording, or otherwise, without the prior written permission of the publisher.

Library of Congress Catalog Card Number 73-138857

07-057349-2

2 3 4 5 6 7 8 9 0   HDBP   7 9 8 7 6 5 4 3

CONTENTS

Preface	vii
1 OBSERVATION AND MEASUREMENT	1
2 THE FUNDAMENTAL PARTICLES	17
3 ATOMS	27
4 IONS	47
5 DISCRETE MOLECULES AND POLYATOMIC IONS	67
6 POLYMERIC MOLECULES AND IONS	91
7 THE PHYSICAL STATES OF MATTER: FLUIDS	119
8 THE PHYSICAL STATES OF MATTER: SOLIDS	145
9 ION EXCHANGE	161
10 PROTON EXCHANGE (ACIDS AND BASES)	179
11 ELECTRON EXCHANGE: OXIDATION AND REDUCTION	197
12 BREAKING AND FORMING COVALENT BONDS	213
13 TRANSFORMATIONS OF MATTER BY THE LOSS OR GAIN OF NUCLEONS	227
Appendix A EXPONENTIAL NOTATION AND THE STANDARD FORM OF A NUMBER	243
Appendix B MEASUREMENT AND UNCERTAINTY	249
Appendix C UNITS OF MEASUREMENT	255
Appendix D CHEMICAL NOMENCLATURE	259
Index	265

PREFACE

This book has been written for the one-semester or two-quarter course in general chemistry. It is assumed that the student has little or no background in chemistry and that he is unlikely to take further work in chemistry, except possibly a one-semester terminal course combining elements of organic chemistry and biochemistry.

On surveying the texts presently available for the terminal chemistry course, we have found two general types. One type retains the historical organization, style, and presentation and the other, while presenting modern chemistry, does so at a level likely to render it incomprehensible to the student with an inadequate background. Our objective, then, was to write a textbook that presents modern chemistry in a way that is comprehensible to the student in the terminal course.

Additionally, we hope to give the student some appreciation of both the methods and the philosophy of science. Fundamental to science is a belief in the rationality of the physical world. Investigations into cause-and-effect relationships of the physical world reveal the order and regularity of that world and have convinced the scientist that it is natural law which functions to produce that order and regularity. A further purpose of this text is to lead students to share this conviction and to value science for what it can do in bringing us to an understanding of our physical world.

The central theme of this book is the structure and transformations of matter. The first eight chapters are devoted to an introductory study of the structure of matter. This study is prefaced by the presentation of the fundamentals of observation and measurement and proceeds to a description of fundamental particles of matter and how they exist in nature. This material continues with a discussion of the chemical bonding of ions and molecules and of the intermolecular and interionic forces in the three states of matter.

The last five chapters are devoted to the transformation of matter, which is presented in terms of the loss and gain of four types of particles—ions, extranuclear protons, electrons, and nucleons—and the making and breaking of covalent bonds.

We have tried to make chemistry relevant to the student by including, where appropriate, discussions of ways in which chemistry affects the student's life and discussion of topics the student is likely to read about in the newspapers—pollution, smoking, and the control of radioactive materials. Although the book contains enough problems to show that chemistry is a quantitative science, arithmetical manipulations are deliberately deemphasized, since we feel that this, more than anything else, "turns off" the nonscience major.

For instructors who feel that more effort should be made in problem solving, the "Student Worktext" accompanying this text may prove helpful. In addition to programmed instruction in equation writing, stoichiometry, and other aspects of problem solving and chemical skills, this "Student Worktext" contains laboratory exercises, exercises with models, and other material to help the student master the basic concepts of chemistry.

B. Richard Siebring
Mary Ellen Schaff

CHAPTER 1 OBSERVATION AND MEASUREMENT

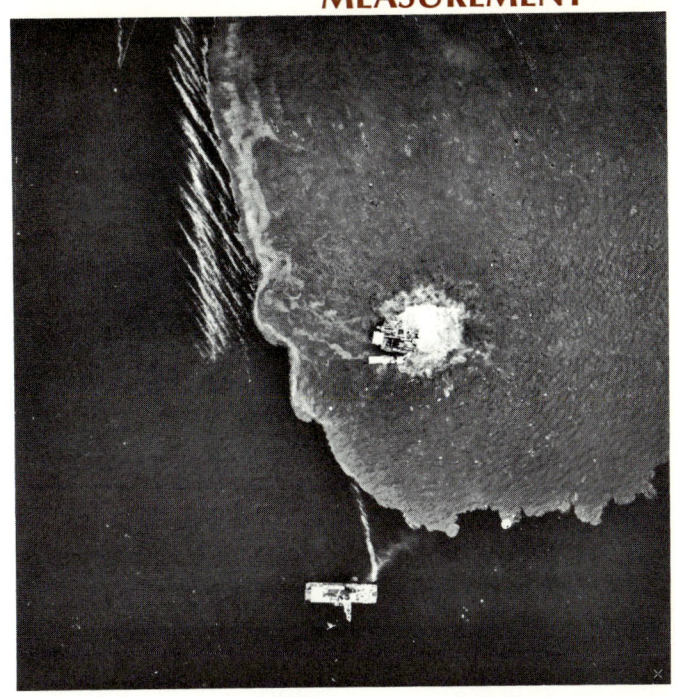

Comparative densities of oil and water are graphically illustrated in this photograph. Less dense than water, oil from spills or seepage floats on the surface of the Gulf, spreading over an ever-widening area and finally washing up on the shore, destroying the recreational properties of the coastline and smothering any wildlife in its path. (Federal Quality Administration and Fish and Wildlife Service, both divisions of the U.S. Department of the Interior.)

CHAPTER 1   OBSERVATION AND MEASUREMENT

Man has always been curious about the materials of his world. That the cave dweller investigated the rocks and soils of his environment is apparent from the colors he used in his cave paintings—blacks, reds, browns, and yellows obtained from natural clays and mineral oxides. His curiosity revealed the comparative strengths of various stones for tools and weapons and later the properties of the earliest known metals, gold, copper, tin, and iron.

Early man may have satisfied his curiosity with the discovery of the utilitarian properties of materials, but as the centuries passed, it is apparent that he became more and more concerned with *understanding* the nature of things. Carried by his curiosity beyond his ability to understand, man turned, in ages past and in primitive societies, to mysticism and magic for explanations of the wonders he observed. As the rationality of the physical world slowly unfolded, however, he became more and more able to hold to an attitude of objectivity when wondering, investigating, and explaining.

Historians place the beginnings of the modern science of chemistry in the seventeenth or eighteenth centuries. To the degree that magic and mysticism were abandoned, the science of chemistry developed. Although the curiosity which impels him is as old as man himself, the attitude which the modern chemist brings to his investigations is new. It is typified by objectivity and rests upon a fundamental belief in the rationality of the physical world. The ultimate purpose of the chemist's activities—and our purpose in this book in considering these activities and their results—is to gain an understanding of the reasons for the properties and behavior of materials.

This first chapter will consider the *properties of matter*, terminology, and some methods and instruments by which these properties can be assessed and described. To relate the properties of materials to their structures requires an understanding of that structure. Accordingly, the next five chapters are devoted to developing such an understanding.

As the story of the structure of matter unfolds, the student must remember that this "story" is a *theory* which chemists have developed to explain the structure of matter. The descriptions of the properties and behavior of matter, on the other hand, are *facts*. When, later in this chapter, a piece of metal is said to weigh 23.4 g, this is the result of direct observation. But when the properties of the electron are described in Chap. 2, they are not and cannot be based upon direct observation. A theory is a generalized and verified statement which explains a group of facts. When the mass of the electron is given as 9.1×10^{-28} g, this is not the result of an actual weighing but a statement that the electron behaves as it would if it did weigh 9.1×10^{-28} g. A theory stands until it encounters facts which contradict it, in which case it must be modified or discarded. It is the essentials of the accepted theory of the structure of matter which Chaps. 2 to 5 will explain.

4 OBSERVATION AND MEASUREMENT

Chapter 2 is concerned with three of the simplest units of structure which chemists describe. The *electron, proton,* and *neutron* are identified and some of their properties are given. Combinations of these simple particles in various proportions are known as *atoms,* the next larger and more complex structural unit. Chapter 3 consists of a detailed examination of the structures of atoms.

The progression from simple to more complex particles leads to an inquiry into the nature of particles formed from atoms. It will be apparent that it is the properties of the atoms themselves that determine whether they form the simple one-atom *ions,* described in Chap. 4, or the polyatomic ions and *molecules,* described in Chaps. 5 and 6.

Atoms, ions, and molecules are associated or clustered in different ways and with different bonding forces. The result is that substances exist in three different *physical states* under different conditions. A close examination of the structural characteristics of matter in each of these three physical states should make the properties of gases, liquids, and solids more understandable (Chaps. 7 and 8).

Chapters 9 to 13 go beyond properties and structure to consider the interactions of materials. These interactions or *reactions,* in addition to being dependent upon the properties and thus upon the structures of the materials reacting, can in many cases be explained as exchanges of one or more of the particles previously studied. Reactions may involve the transfer of ions (Chap. 9), protons (Chap. 10), electrons (Chap. 11), or *nucleons* (Chap. 13). Chapter 12 explains reactions which involve the making and breaking of covalent bonds.

With the close of Chap. 13, the student will have concluded a study of the materials around us which should increase both his understanding of, and his appreciation for, the rationality of the physical world.

For the materials of his world, the scientist uses the term *matter* to designate anything which has mass and occupies space. A second definition necessary for any investigation into the nature of matter is that of *energy.* Energy is defined as the capacity to do work, and it has many forms. The first form of energy with which we shall be concerned will be the most familiar of all—heat energy.

The Properties of Matter

One of the first distinctions we must make about a material is whether it is *homogeneous* (the same throughout) or *heterogeneous.* This is a prime consideration. If a material is to be characterized by the properties we shall describe, there must be an assurance that the material will be the same wherever and whenever it is encountered. This will not be true of some indefinite *mixture* of materials.

The distinction between homogeneous and heterogeneous materials is easier to make in words than to observe in actuality. Some materials are obviously mixtures, such as a piece of granite with its multicolored, particulate makeup, but the additives to table salt which lower moisture retention and reduce lumping are less obvious. The particles of bleaching chemical are differentiated by color in some laundry detergents, but the fact that baking powder is a mixture of several ingredients is less apparent.

The processes by which a chemist can separate and purify the components of a mixture are many and varied. Entire industrial processes may consist of one or more of them. Raw petroleum is a complex mixture of many substances. A large part of petroleum refining is concerned with the separation of these components, or fractions. Air is a mixture of gases. To obtain pure oxygen from the air requires another separatory procedure. Our discussion is less directly concerned with specific techniques of the chemist than with the information gained through those techniques. For our purposes it will be appropriate simply to specify that substances described are homogeneous unless otherwise noted.

We shall describe two classes of homogeneous materials: (1) those with a definite chemical composition, for which we shall reserve the term *pure substances*, and (2) a special class of homogeneous materials, the *solutions*, which may vary in composition; for example, a cupful of water may have either one or two spoonfuls of sugar dissolved in it. Pure substances we shall further divide into those which cannot be decomposed into simpler substances by chemical means (elemental substances or *elements*, such as oxygen, aluminum, sulfur, or lead) and those which can be decomposed by these methods (*compounds*, such as water, carbon dioxide, benzene, and sodium chloride).

OBSERVATIONS OF MATTER

Observations fall naturally into two classes, *qualitative observations* and *quantitative observations*. As the names imply, quantitative observations are numerically expressed, while qualitative observations consist of verbal descriptions. Let us consider some of the *kinds* of observations which can be made about common materials.

Descriptions of stones and minerals are likely to include color, degree of homogeneity, and perhaps heaviness. Descriptions of metals include color, heaviness, luster, and malleability. Liquids such as water, gasoline, or oil may be described in terms of odor, heaviness, or oiliness.

It should be apparent that qualitative observations require verbal precision. Words used must convey identical meanings to the observer and to the people he communicates with. But verbal precision alone will not be enough. It is also essential that there be standards for comparison. In color, for instance, does "red" imply blood red, fire red, or barn red? If we describe a metal as having a luster, is it the dull shine of steel or the polished reflectivity of chromium? The degree of precision and the standards chosen are matters of choice. Sometimes the word "red" is as precise a description as is required, and sometimes the exact wavelength of the red light should be given. In the latter case, however, we have moved out of the area of qualitative observation and into that of quantitative observations, where our descriptions involve numerically expressed measurements.

These *characteristics which identify substances and distinguish them from others* are commonly called *properties*. We are concerned at present only with *physical properties*, by which is usually meant those properties which can be appraised by the senses. We shall see later that materials have another set of properties, *chemical properties*, which can be determined only through interactions (reactions) between materials.

MASS AND WEIGHT

A complete description of an object always requires some assessment of its comparative heaviness. Scientists describe two properties related to the concept of heaviness, properties which may seem at first consideration to be identical. These are *mass* and *weight*. Weight is the gravitational pull or force exerted by the earth on a physical body. That the weight of a body varies with position is an accepted fact in this day of space exploration. The concept of mass is at the same time more fundamental and more difficult to define. Mass is a measure of the quantity of matter in a body, and this is constant. The mass of the astronaut's body has not changed when he walks about on the moon; the apparent lightness of his body is due to the fact that the force of gravity on the moon is less than it is on earth. For earth-bound objects, weight is proportional to mass at any given distance from the center of the earth, and the terms are commonly used interchangeably, or, to be more specific, we refer to the masses of objects as their weights.

As we lift several objects in turn, we may assess their heaviness by comparing them with each other, or we may choose one as a standard and compare the others with it. We can choose the degree of precision, and the more precise it is, the more quantitative the observation becomes. Qualitative observation is the way a scientific observation begins, but if it is to convey exact meanings, it must become quantitative.

MEASUREMENT

Measurement is an everyday operation, but in everyday life the *standard units* are more likely to be the pound, the inch, and the quart of the so-called English system than they are the units of the metric system, the *gram*, the *meter*, and the *liter*. Scientists the world over employ the units of the metric system for measurement. The system has the advantage, in addition to its internationality, of being a decimal system.

7 THE PROPERTIES OF MATTER

A complete table of metric units and English units, together with conversion factors, is given in Appendix C. For our purposes the following units are sufficient, and the student should become familiar with them.

Mass	Length	Volume
Kilogram = 1,000 g	Kilometer = 1,000 m	Kiloliter = 1,000 liters
Gram	Meter	Liter
Centigram = $\frac{1}{100}$ g	Centimeter = $\frac{1}{100}$ m	Centiliter = $\frac{1}{100}$ liter
Milligram = $\frac{1}{1,000}$ g	Millimeter = $\frac{1}{1,000}$ m	Milliliter = $\frac{1}{1,000}$ liter
		(Cubic centimeter = 1.00 cm³ = 1.00 ml)

For comparison: There are 454 g in 1.00 lb.
There are 2.54 cm in 1.00 in.
There is 0.946 liter in 1.00 qt.

These units now serve as our standards for comparison. We need no longer say that one object is "twice as heavy" as another or "almost as heavy" but can say more precisely (quantitatively) that it is as heavy as 35.7 g.

A characteristic property of matter is *density. Density is the weight (mass) of one unit of volume of a substance.* Stated another way, density is the ratio of the weight of an object to its volume. For a solid, in the metric system, density is usually given as the weight in grams of 1.00 cm³ of the substance. Thus a piece of metal which weighs 23.4 g and has a volume of 3.14 cm³ has a density of 23.4 g/3.14 cm³, or 7.45 g/cm³. Densities of solids range from 0.3 g/cm³ for cork and other light woods to 21 g/cm³ for platinum.

To determine the density of a sample, the volume and the weight must be measured, since the density of a sample is found by dividing its weight by its volume. The balances illustrated in Fig. 1-1 serve to illustrate how laboratory tools and instruments are used by the chemist to increase his observational powers and make them more precise. Let us consider the operation of these balances.

    a *The platform balance.* The object to be weighed (say a piece of metal) is placed on the left pan of the balance and known weights are placed on the right pan until the two pans are almost in balance. Final weight is added to the right side by moving the sliding weight, or *rider*, on the arm which is marked off, or *calibrated*, in 1.0- and 0.1-g divisions. For instance, the piece of metal might be exactly balanced by two 10-g weights with the rider at 3.4 g. The weight would be recorded as 23.4 g. We say that the precision of this balance is ±0.1 g, or that it weighs to the nearest 0.1 g.

8 OBSERVATION AND MEASUREMENT

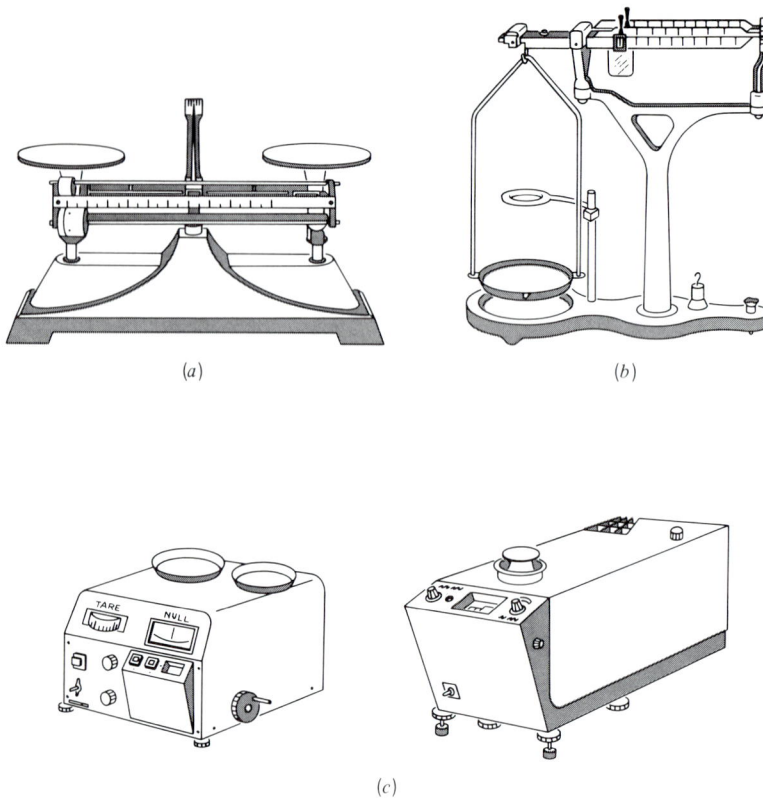

Figure 1-1 Laboratory balances: (*a*) **platform balance,** (*b*) **triple-beam balance,** (*c*) **top-loading balances.**

    b *The triple-beam balance.* The object to be weighed is placed on the pan. The known weights are sliding weights on the three arms of this balance. The middle arm has notches at 10-g divisions, and the back arm has notches at 1-g divisions. The front arm is calibrated in 0.1- and 0.01-g divisions. The piece of metal might be balanced by the weight on the 20-g notch of the middle arm, the weight on the 3-g notch of the rear arm, and the rider on the second division between the 0.4- and 0.5-g markings on the front arm. The weight would be recorded as 23.42 g. We say that the precision of this instrument is ±0.01 g, or that it weighs to the nearest 0.1 g.
    c *Top-loading balances.* These balances, and others of similar design, permit the quick and accurate weighing of an object to a precision of ±0.001 g. The operator controls the balance weights by a system of levers and reads the total weight from an illuminated scale. The piece of metal might have a weight recorded as 23.424 g on this balance.

9 THE PROPERTIES OF MATTER

There are, of course, other instruments for determining weight to even greater precision. As it is with weight, so it is with almost any property of matter: the chemist has many instruments to increase his powers of observation. He selects the instrument with a precision *appropriate* to his task. One might suppose it would always be best to use the instrument with the greatest precision. A simple example should disabuse the reader of this notion. If the metal being weighed sells for 10 cents a gram, and if the balance is to be used solely to determine the price of samples, the price will be exactly the same no matter which of the three balances is used since the price must be given to the nearest cent. In this case it would be difficult to justify the expense of the more precise instrument!

In a determination of density, the volume of the sample must also be found. If the sample is a regular geometric solid, we can measure its dimensions and then calculate its volume. Linear measurement is taken with a meterstick or metric ruler to the nearest 0.001 m or with an instrument called a *vernier caliper* to the nearest 0.0001 m. If the solid is not regular in shape, we may have to use the displacement of water to determine its volume (Fig. 1-2), provided, of course, that the solid is not water-soluble. By dropping the object into a measured volume of water in a graduated cylinder we can determine the increase in the volume of the contents which is due to the presence of the solid.

The volume of a liquid is usually easier to determine than the volume of a solid. When the sample is placed in a graduated cylinder we are able to read the volume in milliliters directly. To determine the density of the sample requires two weighings, however, one of the empty container and the other of the container with the liquid. The difference in the two

Figure 1-2 Determining the volume of an irregular solid by the displacement of water.

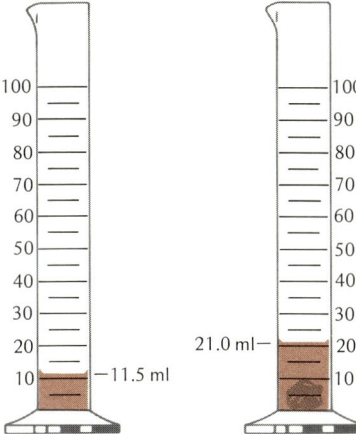

10 OBSERVATION AND MEASUREMENT

weights will be the weight of the liquid sample, and the density will be the weight of the sample divided by its volume.

Specific gravity is a property related to density. Water, the most common of all liquids, is used as a standard of comparison for several properties. The density of water at 3.98°C is 1.00 g/cm³. The specific gravity of a substance is the ratio between its density and the density of water. The ratio will be numerically the same as the density when the density is expressed in grams per cubic centimeter but will have a different numerical value in other units, for instance, in the English system.

If the statement that density (or specific gravity in the metric system) gives the ratio of weight to volume for a sample is properly understood, the student will have no difficulty in calculating the third quantity whenever any two are given.

Example 1-1 What is the density of a liquid sample if its weight is 17.3 g and it has a volume of 19.4 ml?

Solution Since density is the ratio of weight to volume, for this sample it can be expressed numerically as

$$\text{Density} = \frac{\text{weight}}{\text{volume}} = \frac{17.3 \text{ g}}{19.4 \text{ ml}} \text{ or } 0.892 \text{ g/ml}$$

Example 1-2 What will be the weight of 14.3 cm³ of a metal if its density is 7.63 g/cm³?

Solution Density, being the ratio of weight to volume, gives directly the weight of a unit of volume. Therefore if 1.00 cm³ of the metal weighs 7.63 g, 14.3 cm³ of the same metal will weigh 14.3 times as much, or

$$(14.3 \text{ cm}^3)(7.63 \text{ g/cm}^3) = 109 \text{ g}$$

Example 1-3 A sample of copper shot weighs 104 g. Copper has a density of 8.93 g/cm³. What volume will this sample displace if it is dropped into a graduated cylinder partially filled with water?

Solution The density, 8.93 g/cm³, is the weight of 1.00 cm³ of copper. Then 104 g must be the weight of several cubic centimeters. To find how many we divide

$$\frac{104 \text{ g}}{8.93 \text{ g/cm}^3} = 11.6 \text{ cm}^3$$

The Three States of Matter We may group or classify substances on the basis of any one of the several properties we have already mentioned—color, odor, or luster or the lack of it. A classification more satisfactory than any of these is to group

11 HEAT EFFECTS

substances according to whether they are *solid, liquid,* or *gas* under ordinary conditions. These three terms are used to describe the *physical states* of matter. In later chapters of the book we shall examine the nature of each of these states in greater detail. For the time being we shall use them simply as convenient classifications. Let us next consider which properties are especially appropriate for descriptions of matter in each of the three states.

SOLIDS

In addition to *appearance* and *density,* already discussed, some experimentation will be required to determine such things as *hardness, brittleness, malleability,* and *elasticity.* A qualitative impression of these properties can be obtained quite easily. The piece of iron has a hardness far exceeding that of the piece of lead. The brittleness of glass as compared to that of some plastics has led to the development of plastic windshields. The malleability of gold, and to a lesser extent of silver and copper, gave them importance in early history, not only for their decorative effects but because they could be worked and shaped. Although elasticity is associated in our minds with rubber and other stretchables, it also is a property which must be considered when selecting metals for construction.

LIQUIDS

Color and *transparence* are important properties of liquids, as is density. The variation in density of two liquids has been demonstrated all too graphically and too often as oil—less dense than water and thus floating on top—washes upon our beaches.

Viscosity, sometimes termed resistance to flow, is a highly characteristic liquid property and one which can be measured precisely. Even simple qualitative observations distinguish between the viscosities of such easily flowing liquids as gasoline or water and the more viscous lubricating oils, molasses, or tar.

GASES

Of all properties of gases, we are probably most impressed by *odor.* It is a highly characteristic property, although it will remain for us a qualitative determination. Although *density* is a characteristic property, it is not as easily determined for gases as for liquids or solids. Densities of gases are given as weights of a liter and range from 0.08 g/liter for the lightest gas, hydrogen, to almost 10 g/liter for radon, the heaviest gas.

Heat Effects As long as our search for the characteristic properties of substances is limited to physical properties, we are not concerned with interactions between substances. However, when listing properties of solids, we were concerned with reactions to physical stress, such as bending, scratching, or stretching. Another type of physical strain can be placed upon substances, and reactions to it still come under the heading of physical properties. This strain is the application of *heat energy*. When we say that a material has been heated, this does *not* mean that it has been ignited or caused to burn but simply that its temperature has been increased. A selection of solid samples can be placed upon a metal plate and then the heat of a bunsen flame directed to the plate from beneath. It will soon be apparent that solids react differently to heat. Everyone knows what happens to a block of ice subjected to the application of heat but perhaps has less idea of the relative effects of heat upon such substances as aluminum, tallow, lead, and sulfur. *Melting* will be found to be a characteristic property of solids, a property given a quantitative meaning if we record the temperature (*melting point*) at which it occurs.

Liquids likewise have characteristic behavior when heated. All liquids can be made to *boil* or vaporize if heated to high enough temperatures. This property also becomes quantitative with the recording of the temperature (*boiling point*).

To record temperatures a thermometer is used. Here, again, the standard unit of the scientist is different from that used in everyday circumstances. We note the outdoor temperature or the fever of an illness on a thermometer whose unit is the Fahrenheit degree. On this thermometer the freezing point of water is 32° and the boiling point of water is 212°. The Celsius scale of the scientist is simpler. On this scale the freezing point of water is 0.0°, and the boiling point is 100°.

In Fig. 1-3, a Celsius thermometer and a Fahrenheit thermometer are drawn side by side in such a way that the melting and boiling points of water are at the same relative positions. As can be seen, the Celsius degree is larger than the Fahrenheit degree since it takes 180 Fahrenheit degrees to measure the temperature difference which only 100 Celsius degrees measure. In fact, the Celsius degree is exactly 180/100, or $\frac{9}{5}$, larger than the Fahrenheit degree. The chemist deals almost exclusively with the Celsius scale, so that conversions from one system to the other are not often necessary. Formulas for these conversions are given in Appendix C.

The boiling points and melting points of substances are temperatures and are measured in degrees. Temperature is a measure of *heat intensity*. Another aspect of heat is *quantity of heat*. Consider two identical beakers, one containing 100 g of water, the other containing 200 g of water. If each in turn is heated to boiling by the same bunsen burner, held at the same respective position, it will take twice as long to heat the larger sample to the boiling point as it did the smaller. The larger amount of water requires the larger amount of heat. The unit used to measure the

13 HEAT EFFECTS

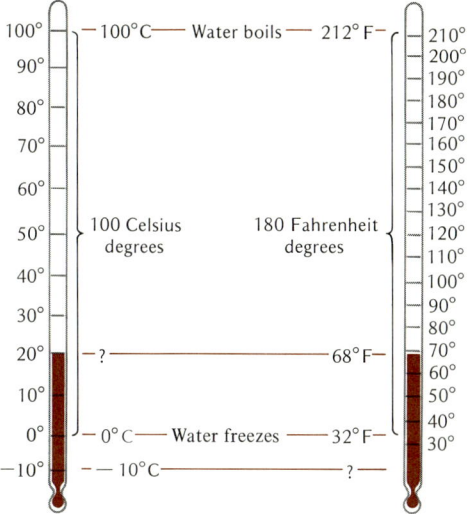

A Celsius thermometer   A Fahrenheit thermometer

Figure 1-3 Thermometers.

quantity of heat is the *calorie*. The calorie is defined as that amount of heat which will raise the temperature of one gram of water one degree Celsius. This unit is also referred to as the *small calorie*. The *large calorie* of the nutritionist is 1000 small calories.

Example 1-4   How many calories are required to heat 200 g of water from a room temperature of 27°C to boiling?

Solution   It takes one calorie to raise one gram of water through one degree in temperature. Therefore 200 g would require 200 times as many calories, and because the temperature change is to be through 73 degrees, it will require 73 more times as many calories:

$$(200 \text{ g})(73°C)[1.0 \text{ cal}/(g)(°C)] = 14{,}600 \text{ cal}$$

Another interesting heat effect can be demonstrated if two identical beakers, one containing 100 g of water and the other containing 100 g of carbon tetrachloride, are heated by identical burner flames for the same period of time. Thermometers placed in the liquids will show that at the end of any given time, the carbon tetrachloride will have reached a higher temperature than the water. It takes a larger quantity of heat to change the temperature of water than to change the temperature of almost any other common substance. The quantity of heat necessary to bring about a temperature change of one degree Celsius in a one-gram sample of a substance is called the *specific heat*. Metals have very low specific heats; for instance aluminum has a specific heat of 0.212 cal/(g)(°C). This means that a given quantity of heat will have almost 5 times the effect upon the temperature of 1 g of aluminum as it will upon the temperature of 1 g of water, or that a given quantity of heat can raise the temperature of almost 5 times as much aluminum as water to the same temperature. The specific heat of a substance is a characteristic property.

Chemical Properties All observations of the nature of materials described so far can be made without fundamentally altering the material. To melt a block of ice may seem a fundamental change, but as soon as the heat necessary to its melting is withdrawn, we obtain the ice again with all its properties intact. Smashing a stone does not alter the material; it is still the same material, in a smaller state of subdivision. All physical properties can be observed with no more than *physical changes* being effected in the materials. Every substance has another set of properties, known as *chemical properties*. These properties are observable only through interactions between substances. These interactions are called *chemical changes*. When a substance undergoes a chemical change, it becomes a different substance with a distinctly different set of properties. Some common chemical changes are burning, rusting, and fermentation. We shall postpone our examination of chemical reactions until a later chapter.

Summary We have made a beginning in our investigation into the nature of substances. Several physical properties especially useful in characterizing and classifying substances have been found to be:

    1 Classification as solid, liquid, or gas
    2 Qualitative determinations
       a Appearance
       b Odor
       c Reaction to physical stress (solids)
       d Resistance to flow (liquids)
    3 Quantitative determinations
       a Density
       b Melting point (solids)
       c Boiling point (liquids)
       d Specific heat

Our observations are not intended to be an end in themselves. What we are really seeking, of course, is an explanation for *why* one material exists as a gas while another is a liquid; why a metal is shiny while sulfur

15 QUESTIONS

is dull; or what accounts for the differences in density among substances. Before we can obtain these explanations, we must consider the structure of materials in greater detail. This will be our concern in the next seven chapters. Unlike the observation of common properties, however, knowledge of the submicroscopic structure of materials is based upon very sophisticated measurements.

Questions

1 What is meant by homogeneous? What kinds of matter are homogeneous?

2 Is a mixture necessarily heterogeneous? Explain.

3 What is another term for homogeneous mixture?

4 Consider objects and materials with which you are familiar. Name several solids which appear to be homogeneous throughout and several which appear to be heterogeneous. Do the same with familiar liquids. Can you do the same with gases?

5 Are most foods homogeneous or heterogeneous? Name some homogeneous foods.

6 Consider the following common materials and list the *kinds* of qualitative observations which might be made about each:

Vinegar   Chalk
Salt   Copper wire
Lubricating oil Air
Charcoal   Milk

7 What are two types of pure substances? How do they differ from each other?

8 Will an object weigh more or less at the top of a mountain than at the bottom? How will its mass vary under these conditions?

9 Why is a triple-beam balance sometimes referred to as a "centigram balance"?

10 What misunderstanding is taken advantage of in the question: "Which is heavier, a pound of lead or a pound of feathers?"

11 Is fuel oil more or less dense than water? What evidence can you cite for your answer? In what other very noticeable property do these two liquids differ?

12 A cube of unknown material is found to measure 2.00 cm on a side, and to weigh 54.0 g.
  a What is the volume of the cube in cubic centimeters?
  b Calculate the density of the cube in grams per cubic centimeter.
  c What is the specific gravity of the cube?

13 Metal filings were added to a graduated cylinder containing 10.3 ml of water until the total volume reached 28.9 ml. What was the volume of metal filings added?

14 A graduated cylinder containing an unknown liquid gave a volume

16 OBSERVATION AND MEASUREMENT

reading of 13.2 ml and was found to weigh 54.8 g. When the liquid was poured out and the cylinder reweighed, the weight was 43.7 g.
 a What was the weight of the liquid in grams?
 b Calculate the density of the liquid in grams per milliliter.
15 An object with an irregular shape was weighed. Its weight was found to be 20.4 g. When placed in a graduated cylinder containing 14.3 ml of water so that the object was completely submerged, it caused the water level to rise to 17.1 ml.
 a What was the volume of the object in milliliters? (Assume that it was completely water-insoluble.)
 b What was its volume in cubic centimeters?
 c Calculate the density of the object.
16 What is the weight of 1.00 liter of water at 3.98°C?
17 Sulfuric acid has a density of 1.8 g/ml. What is the weight of 1.00 liter of sulfuric acid?
18 The specific gravity of a liquid is 1.30. What is the weight of 513 ml of this liquid in grams?
19 How many calories will it take to raise the temperature of 500 g of water from room temperature (27°C) to the boiling point?
20 What distinguishes physical properties from chemical properties?
21 Classify the following as physical or chemical properties.
 a Odor   b Color
 c Taste   d Electric conductivity
 e Density f Flammability
 g Flexibility h Viscosity
22 What are four physical properties of copper?
23 Which of the following properties of iron are properly classified as chemical properties?
 a Silver-white in color
 b Corrodes to rust in the atmosphere
 c Density of 7.86 g/cm^3
 d Melts at 1535°C
 e Boils at 2887°C
 f Reacts with steam at high temperatures to form a compound sometimes called magnetic oxide
 g Attracted to a magnet
24 Distinguish between members of each of the following pairs of terms:
 a Compound and element
 b Compound and solution
 c Weight and mass
 d Qualitative and quantitative observations
 e Pure substance and mixture
 f Specific gravity and density
 g Temperature and amount of heat

CHAPTER 2   THE FUNDAMENTAL PARTICLES

Lightning resembles the cathode rays discussed in this chapter in that it is a discharge of electricity through a gas. Both phenomena result from a buildup of a charge between two objects connected by a gas only. (W. Prapst and Agfa-Gevaert-Bildarchw.)

CHAPTER 2 THE FUNDAMENTAL PARTICLES

One of the questions considered by the early Greek philosophers, who were singularly attracted to the fundamental mysteries of nature, was the question of whether matter was continuous and infinitely divisible or was discontinuous, consisting of a multitude of separate, indivisible pieces. Leucippus (5th century, B.C.) and his student Democritus (470-380 B.C.) argued the case for discontinuity. They believed that there were ultimate particles beyond which further division was impossible and which were surrounded by empty space. Aristotle (384-322 B.C.) and other influential philosophers could not accept the idea that there were particles of matter which could not be split into still smaller pieces and hence argued that matter was infinitely divisible and continuous. The latter point of view prevailed, and matter was regarded as continuous for over 2,000 years. Then, in 1805, John Dalton (1766-1844) revived the concept of ultimate particles of matter.

With the Greeks, the idea of the discontinuity of matter was purely a philosophical concept; they did not attempt to support their conclusions with experimentation. Dalton's concept of ultimate particles, however, while not built on an experimental foundation, was widely accepted because he used it to explain certain chemical principles which had been established through experimentation. At present, the concept of the discontinuity of matter and the existence of ultimate indivisible particles is supported by considerable experimental evidence. It should be noted, however, that the fundamental particles as conceived by the modern scientist are different from what Dalton considered the ultimate particles of matter. What Dalton believed to be the ultimate particles have actually turned out to be highly divisible complex bodies (see Chap. 3).

How many kinds or types of fundamental particles are there? The answer to that question depends upon what authority one refers to. Some scientists talk of 100 different types of fundamental particles. Others are convinced that nature is not that complicated and view matter as being composed of a much smaller number of particles. To understand ordinary chemistry, we need concern ourselves with only three fundamental particles—electrons, protons, and neutrons. This chapter is devoted to a description of these particles and some of the experimental evidence which supports this description.

Electricity Since two of the three fundamental particles are electrically charged, it is helpful at this point to discuss the nature of the phenomenon we call electricity.

Even though the early Greeks were men of thought rather than experimentation, they observed as early as 500 B.C. that a piece of amber which had been rubbed with fur was capable of picking up small pieces of cloth, lint, or other materials by attracting these light objects to itself. In the sixteenth century this property was given the name *electricity*, after the Greek word *elektron*, meaning amber. It was observed at about the same time that electricity is not exclusively a property of amber but that certain other substances, such as glass, also acquire electricity when rubbed with certain kinds of cloth.

In the eighteenth century it was learned that there are two kinds of electrification. In one case electrified objects repel each other; in other cases they attract each other. For example, two electrified amber rods (or two electrified glass rods) repel each other, but an electrified amber rod and an electrified glass rod attract each other. The type of electricity acquired by glass became known as positive electricity, and the kind acquired by amber as negative electricity.

If a positively charged object and a negatively charged object are brought in contact with each other, they lose their electricity. We *say* that electricity flows from the positive object to the negative object and they neutralize each other. However, there is no scientific basis for using these terms; they are merely conventions which originated when Benjamin Franklin, in his early investigations of electrical phenomena, arbitrarily designated electrified glass as positive. Positive and negative objects also lose their electricity if they are not allowed to touch each other but are connected by a copper wire. In this case, we say that electricity flows from the positive object through the wire to the negative object. The flow of electricity through the wire is referred to as an *electric current*.

Many units, with very precise definitions, are used to describe an electric current. We need consider only a few of them. Our discussion will be limited to a description of what they measure and will not involve precise definitions.

One way of making these units meaningful is to compare them to the units used to measure the flow of water. The quantity of water flowing past a given point in a pipe might be measured in milliliters. An analogous unit used to measure the quantity of electricity flowing past a given point in a wire is the *coulomb*. The rate of flow of water in the pipe might be milliliters per second. The rate of flow of electricity in a wire is measured in coulombs per second, or *amperes*. Finally, the rate of flow of water through a pipe is determined by the difference in the pressure at the ends of the pipe and is measured in such units as grams per square centimeter. The rate of flow of electricity is determined by the difference in the intensity of the electric charges at each end of the conductor. This is referred to as potential difference and is often measured in *volts*.

We are accustomed to thinking of electricity as passing readily through certain solid objects, but electricity will also pass through gases and some liquids. For example, if positively charged and negatively charged objects are placed sufficiently close to each other, and if the potential difference

21 ELECTRICITY

is sufficiently large, electricity will pass between them in the form of a spark. Lightning is an example of the passage of electricity through a gas. One of the fundamental particles, the electron, was discovered as a result of studying the passage of electricity through gases.

THE ELECTRON

Let us consider a glass tube, containing a gas, about 1 ft long and 1 in. in diameter, equipped with metal disks sealed in each end and an outlet connected to a vacuum pump so that the gas can be removed from the tube. Such a tube is shown in Fig. 2-1. Each metal disk is connected to an outside source of an electric current. When the electric current is first turned on, nothing much is observed unless extremely high voltages are used, but when most of the gas is removed so that a near vacuum is created, the current will flow more readily. This flow of an electric current through a gas at very low pressure results in a luminosity in the tube. The early observers of this phenomenon correctly concluded that this luminosity was caused by *rays* originating at the negatively charged disk (the cathode), and consequently they referred to them as *cathode rays*.

Investigations of the behavior of these rays have revealed the following properties:

1. They travel in straight lines. This is shown by the fact that a metal object placed in the path of the rays will cast a distinct shadow (see Fig. 2-2).
2. They evidently travel with considerable energy, since they will rotate a small paddle wheel placed in their path (see Fig. 2-3).
3. They are deflected by a magnetic field in a direction that would indicate that they consist of negatively charged particles (see Fig. 2-4).
4. The fact that they are attracted by positively charged objects outside

Figure 2-1 Cathode-ray tube.

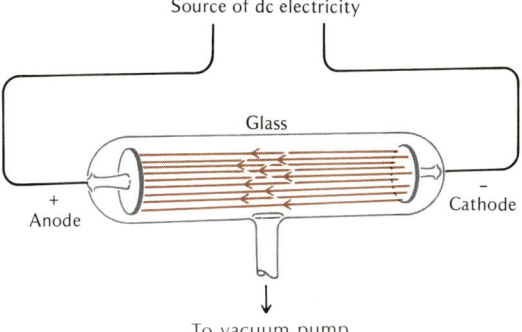

22 THE FUNDAMENTAL PARTICLES

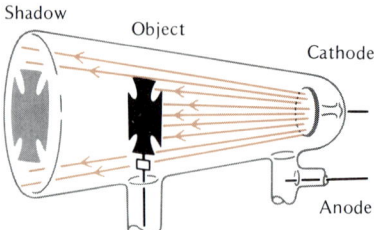

Figure 2-2 Cathode-ray tube with an obstruction in the path of the rays.

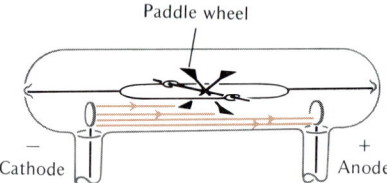

Figure 2-3 Cathode-ray tube with a small paddle wheel in the path of the rays.

the tube also indicates that the particles which make up the cathode rays must be negatively charged (see Fig. 2-5).

Intensive investigation of these particles has shown them to be extremely small. The mass of a single particle is only 9.1×10^{-28} g. This is the smallest mass of any known particle of matter. In addition, the cathode-ray particles carry the smallest quantity of charge that has ever been measured; no particle has been found carrying a smaller charge. These particles are now called *electrons*, and their charge is referred to as the *unit charge* (-1).

The nature or behavior of cathode rays is independent of the materials in the tube, the metal in the electrodes, or the gas in the tube. It was soon learned that there were sources other than the cathode-ray tubes for electrons. Thomas A. Edison (1847-1931) showed that electrons are emitted by metals heated to a high temperature in a vacuum. J. J. Thomson (1856-1940) discovered that electrons can be removed from metals by irradiating their surfaces with ultraviolet light. In every case, electrons appear to be identical. Apparently it is possible to remove electrons from many ordinary materials. When we walk across a carpet on a dry day, we remove electrons from it which accumulate on the body surface. When we come close to something metallic, these electrons jump across the gap, creating a spark. Evidently electrons are everywhere in all forms of matter.

ELECTRICITY

Man has learned to control the behavior of electrons in matter and space. This has resulted in the vast electrical and electronic industries. The reader is well aware of the tremendous number of uses of an electric current in our society. The picture tube in a modern television set is a modified cathode-ray tube. The photoelectric cell used to open a garage door uses the electrons that are emitted when ultraviolet light strikes a metal surface. Vacuum tubes in radios operate because electrons are emitted when metals are heated to a high temperature in a vacuum.

We now know that negatively charged objects, such as the fur-rubbed amber, have an excess of electrons; and conversely, the fur is deficient in electrons and is positively charged. Similarly, the cloth-rubbed glass acquires a positive charge by transferring electrons to the cloth, which becomes negatively charged. The electric current produced when a negatively charged object is connected by a wire to a positively charged object is a movement of electrons in the wire. That is, electrons are taken from the surface of the negatively charged object and transferred to the wire, and electrons are also taken from the wire and transferred to the surface of the positively charged object. The reader should note that the electron flow is opposite to the *conventional* flow of positive electricity as mentioned on page 20. This unfortunate situation is the result of the arbitrary designation of electrified glass as being positively charged. If the opposite choice had been made, the electron would have been positively charged and this anomalous situation could have been avoided. Unfortunately, science is now so imbued with special affiliations of positive and negative charges that it seems impossible to make any change.

Figure 2-4 Cathode-ray tube in a magnetic field.

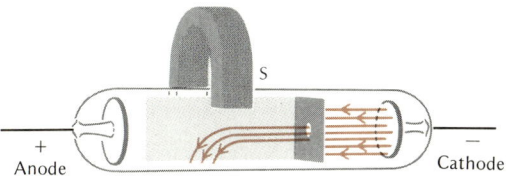

Figure 2-5 Cathode-ray tube in an electric field.

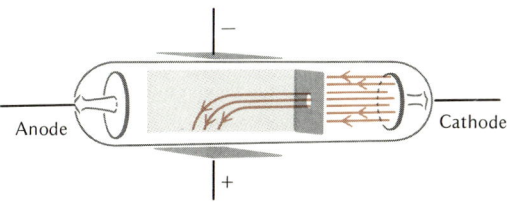

THE PROTON

Now let us turn our attention to another type of cathode-ray tube, such as the one diagrammed in Fig. 2-6. This tube differs from the one previously discussed in that the cathode is placed further up in the tube and is perforated. Early investigators showed that in such a tube another stream of particles forms behind the cathode. Like electrons, these particles travel in straight lines and are deflected by a magnet. However, the particles were quickly shown to be something other than electrons, since they are attracted by negatively charged objects. These particles are positively charged and much heavier than electrons. Furthermore, both the mass and charge on these particles vary, depending on the residual gas in the tube. The masses of these particles were shown to be nearly the same as the mass of the particles which made up the gas in the tube. These observations led the experimenters to the conclusion that the origin of these particles is the residual gas and not the anode.

We now believe that these positive particles are formed when electrons emitted from the cathode strike neutral gaseous particles. When they strike, additional electrons are removed from the neutral particle and positively charged particles of very nearly the same mass as the neutral particle are formed. The positive particles thus formed are attracted to the cathode, and if they have sufficient velocity upon reaching the vicinity of the cathode, they will pass through the holes in the cathode and form the rays, which are often referred to as the *positive* or *canal rays*.

The smallest particle found in positive rays was that formed when hydrogen gas was used in the tube. This particle is referred to as the *proton* and has a mass 1,840 times that of an electron. The proton has a charge equal but opposite to that of an electron (+1). We now accept the proton along with the electron as one of the three fundamental particles.

THE NEUTRON

The neutron is the third fundamental particle. It is neutral and has approximately the same mass as a proton. Unlike the proton and electron, the isolated neutron spontaneously disintegrates. The products of this disintegration are a proton and an electron. For this reason, some authorities regard the neutron as a coalescence of a proton and an electron. The

Figure 2-6 Cathode-ray tube with perforated cathode.

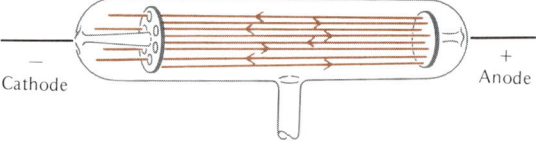

TABLE 2-1 Comparison of Negatively and Positively Charged Particles Formed in Cathode-ray Tubes

Charge	Origin	Properties
Negative	Cathode	Characteristics are always the same—independent of the materials in the tube, electrodes, and residual gas
Positive	Residual gas present in the tube	Varies with the gas in the tube

TABLE 2-2 Comparison of Fundamental Particles

Name	Charge	Mass
Electron	Smallest quantity of charge ever measured, now referred to as the unit charge (-1)	Smallest mass of any known particle, 9.1×10^{-28} g
Proton	Same quantity of charge as on electron but opposite in sign $(+1)$	About 1,840 times that of an electron, 1.67×10^{-24} g
Neutron	Neutral	Approximately the same as that of a proton

experimental work which led to the discovery of the neutron will be discussed in a later chapter.

As we have indicated previously, there are scores of other fundamental particles described by physicists, but it is possible to explain the myriad kinds of matter in terms of only the three fundamental particles presented here. The properties of these particles are summarized in Table 2-2.

In the next chapter we shall see that electrons, protons, and neutrons are assembled into a relatively small number of combinations which we call atoms. In subsequent chapters we shall see that these assemblages are in turn combined to form many more complex building blocks of matter.

Questions

1 What difference in Dalton's theory as compared to the atomic theory of Democritus probably brought it the support and acceptance that the earlier theory was unable to command?

2 Explain electric charge and electric current in terms of fundamental particles.

THE FUNDAMENTAL PARTICLES

3 What is meant by the statement that the "conventional" flow of electricity is opposite in direction to that of the electron flow?
4 Can you offer any experimental evidence to support the following statements?
   a All forms of matter contain electrons.
   b Cathode rays travel in straight lines.
   c Electrons are negatively charged.
   d Electrons consist of particles of matter rather than simply radiant energy.
5 What carries the current in a copper wire? A cathode-ray tube?
6 Why do some scientists not regard a neutron as a fundamental particle?
7 Compare the proton, neutron, and electron with respect to mass and charge.
8 How can a given quantity of matter be neutral if it contains charged particles?
9 What is the origin of cathode rays? Positive rays?
10 Classify the following statements as (1) true for cathode rays only, (2) positive rays only, (3) both cathode and positive rays, (4) neither cathode rays nor positive rays.
   a They travel in straight lines.
   b They consist of charged particles.
   c They are deflected by a magnet.
   d They are attracted by a negative field.
   e They consist of negatively charged particles.
   f They originate at the cathode.
   g They originate at the anode.
   h Their mass varies with the kind of gas in the tube.
   i Their mass is independent of the kind of matter the tube is made of and of the kind of matter inside the tube.
11 Calculate the mass of the proton from the mass of the electron, 9.1×10^{-28} g, and the fact that the mass of the proton is 1,840 times that of the electron.

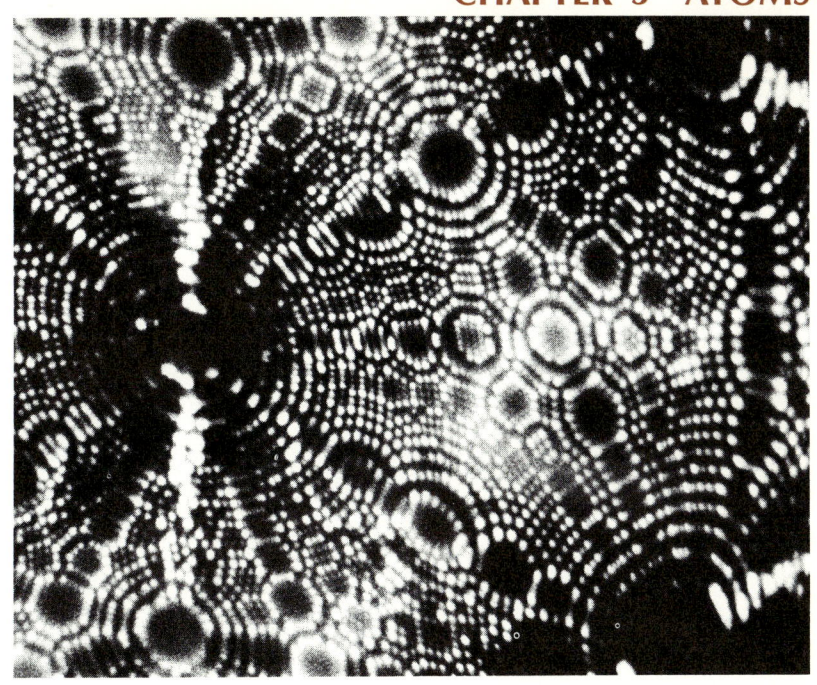

CHAPTER 3 ATOMS

With invention of the field-ion microscope by Professor Erwin W. Müller in 1951 (now at Pennsylvania State University), the atom has ceased to be an entirely theoretical concept. This instrument permits one to locate individual atoms in a metal surface. Each bright spot on this photograph represents an atom in the surface of a tungsten needle. This is magnified approximately 3 million times. (Professor Ralf Vanselow, Chemistry Department, University of Wisconsin, Milwaukee.)

CHAPTER 3 ATOMS

Fundamental particles exist in nature in clusters known as atoms. Over 1,400 different kinds of atoms, or clusters of fundamental particles, are known. However, only 325 occur naturally; the others are man-made. Of the 325 naturally occurring atoms, 260 are stable and 65 are unstable. Unstable atoms disintegrate spontaneously and are referred to as *radioactive atoms*. This disintegration of a radioactive atom usually takes place by the emission of electrons, called *beta particles*, and/or clusters of two neutrons and two protons, called *alpha particles*.

Atom is derived from the Greek word *atomos*, which means indivisible. Democritus and other Greek philosophers who argued for the discontinuity of matter used this word to refer to the ultimate particle of matter. The word "atom" was also adopted by the English chemist John Dalton, who supposed atoms to be indestructible particles having a characteristic weight for each element. He thought of atoms of the same element as being alike in all respects and atoms of different elements as being different, especially in weight. Because of his contributions to the concept of the atom, Dalton is sometimes referred to as the father of the atomic theory.

The atom is now defined as the smallest particle of an element, with each element having its own characteristic atoms. However, we no longer conceive of atoms as indivisible bodies but believe that they are highly complex clusters of fundamental particles.

The Nuclear Atom

The concept of the nuclear atom originated as a result of an experiment performed in 1911 by Ernest Rutherford (1871–1937) on the penetrability of matter by alpha particles. An alpha particle, which consists of two protons and two neutrons, is emitted by naturally radioactive elements. It can also be obtained by knocking two electrons off an atom of the gas helium in a cathode-ray tube.

Rutherford used polonium, a radioactive element, as a source of alpha particles. He arranged to have a narrow stream of alpha particles "shot" at gold foil about 0.0005 in. thick. This was accomplished by placing a small quantity of polonium in a deep hole of small diameter bored into a thick block of lead. Several sheets of heavy lead foil, each perforated with a tiny hole, were placed between the source of alpha particles and the gold foil. These holes were aligned with the hole in the block of lead. Since thick pieces of lead absorb alpha particles, only those alpha particles which were traveling in a specific straight line were allowed to strike the gold foil (see Fig. 3-1).

One reason why Rutherford selected alpha particles for this experiment was that they are easily detected. His method of detection used the same principle as that employed in luminescent clocks. Certain materials, called phosphors, scintillate or sparkle when struck by alpha particles. Metal

30 ATOMS

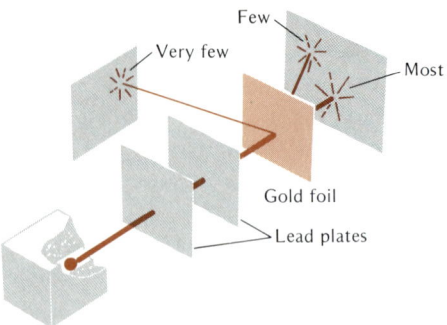

Figure 3-1 Diagram of Rutherford's experiment.

sheets covered with zinc sulfide, a phosphor, were arranged in such a manner that alpha particles which passed through the gold foil or were deflected by it could be detected.

Rutherford reasoned that since gold cannot be compressed significantly even by tremendous pressure, gold atoms must be adjacent to each other, and there is very little space, if any, between them. Hence, if any alpha particles passed through the gold foil, they would be passing through the atoms and not between them. Actually, it was observed that the vast majority of the alpha particles passed through the foil as if it were not there.

Since the alpha particles passed through several thousands of atoms, this was a rather startling observation. One could only conclude that the gold atoms were essentially space.

Where was the mass of the gold foil? The answer to this question lay with the 1 alpha particle in 100,000 which was deflected. A large fraction of the alpha particles which were deflected were deflected by very large angles. A few were even deflected back in the direction from which they came (see Fig. 3-2). These observations led Rutherford to the conclusion that the mass of the atom is concentrated in a very small portion of the

Figure 3-2 Bombardment of gold foil with alpha particles.

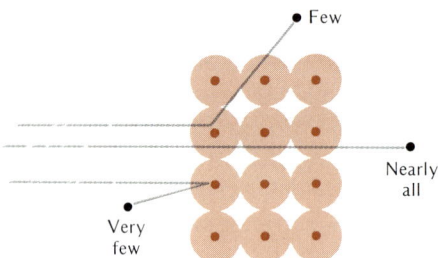

total volume of the atom, which he called the *atomic nucleus*. He calculated the volume of the nucleus to be only about 10^{-15} times the volume of the atom. By analyzing the nature of the collisions between alpha particles and gold nuclei, Rutherford concluded that the nucleus of an atom is positively charged. It was postulated further that the nucleus contains the protons of the atom and that the electrons are outside the nucleus. (Neutrons, which were not identified until 1932, are also found in the nucleus.) In a neutral atom the number of protons inside the nucleus is the same as the number of electrons outside the nucleus. This number is called the *atomic number*.

It is the atomic number which identifies an atom of a given element. Contrary to the postulates of John Dalton, we now know that atoms of the same element do not necessarily have the same mass. They do, however, have the same number of protons in the nucleus, or the same atomic number. There are 105 elements, with atomic numbers of 1 to 105. Only 89 elements are naturally occurring. The remainder are man-made.

If the atoms of a given element have the same atomic number but may have different atomic mass, it follows that atoms of this element must differ in the number of neutrons in the nucleus. Atoms of the same element (and hence the same atomic number) with different numbers of neutrons are referred to as isotopes. Atoms of an isotope of a given element not only have the same atomic number but also the same *mass number*, the number of *nucleons* in the nucleus. (The term nucleon is used to refer collectively to protons and neutrons.)

Oxygen, the most abundant element, has three naturally occurring isotopes. The most abundant of these (99.76 percent) has a mass number of 16. The other two natural isotopes have mass numbers of 17 and 18. All atoms of this element have eight protons in their nuclei and, if the atom is neutral, eight electrons in the remainder of the atom. The isotope with a mass number of 16 has eight neutrons in the nucleus. The isotopes with the mass numbers of 17 and 18 have nine and ten neutrons respectively in the nucleus. The structures of the atoms of some other naturally occurring isotopes are presented in Table 3-1.

Atomic Weights

In addition to the mass number and atomic number, another number, atomic weight, is often used to describe atoms. Although it is now possible to determine the weight of a single atom with a high degree of accuracy, it is much more convenient to refer to weights of atoms in terms of a ratio. This ratio expresses the weights of all atoms relative to a common standard and is called *atomic weight*. This standard has changed a number of times. In 1961 the International Union of Pure and Applied Chemistry designated the most common isotope of carbon (mass number 12) as the standard and assigned it the atomic weight of 12 *exactly*. This

TABLE 3-1 Structures of Some Naturally Occurring Isotopes

Element	At. No.	Mass No.	Nucleus Protons	Nucleus Neutrons	Charge	Electrons in Neutral Atom	Abundance, %	Atomic Weight Of Isotopes	Atomic Weight Of Element
Hydrogen	1	1	1	0	+1	1	99.984	1.008	1.008
		2	1	1	+1	1	0.016	2.014	
Carbon	6	12	6	6	+6	6	98.9	12.000	12.011
		13	6	7	+6	6	1.1	13.003	
Nitrogen	7	14	7	7	+7	7	99.62	14.003	14.007
		15	7	8	+7	7	0.38	15.000	
Magnesium	12	24	12	12	+12	12	78.60	23.985	24.312
		25	12	13	+12	12	10.11	24.986	
		26	12	14	+12	12	11.29	25.983	
Chlorine	17	35	17	18	+17	17	75.4	34.969	35.453
		37	17	20	+17	17	24.6	36.966	
Uranium	92	234	92	142	+92	92	0.01	234.04	238.03
		235	92	143	+92	92	0.71	235.04	
		238	92	146	+92	92	99.28	238.05	

assignment was arbitrary, but once it was established, other atoms could be compared accurately to it. Thus to say that a given atom has an atomic weight of 24.0 means that it is twice as heavy as the atom of carbon which has a mass number of 12. Similarly an atom with an atomic weight of 6.0 is half as heavy as the standard. The atomic weights of both protons and neutrons are nearly (but not exactly) equal to 1.

When discussing the atomic weight of an element, we are referring to the *weighted* average of the atomic weights of the isotopes which make up the element. For example, the element magnesium consists of three naturally occurring isotopes with atomic weights of 23.985, 24.986, and 25.983. These isotopes occur in nature in such proportion that the *weighted* average atomic weight for magnesium is 24.312.

A convenient unit for measuring the quantity of an element is the number of grams equal to its atomic weight. This unit is referred to as the *gram atomic weight*. This unit is convenient because gram atomic weights are in the same ratio as atomic weights, and hence 1 gram atomic weight of an element must contain the same number of atoms as 1 gram atomic weight of any other element.

Symbols Dalton used a symbol to represent each element. His symbols consisted of various marks or designs inside of circles (see Fig. 3-3). For example, a circle with a central dot represented hydrogen; a solidly shaded circle

33 THE ELECTRON THEORY OF THE ATOM

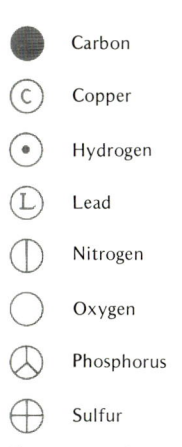

- ● Carbon
- Ⓒ Copper
- ⊙ Hydrogen
- Ⓛ Lead
- Nitrogen
- ○ Oxygen
- Phosphorus
- ⊕ Sulfur

Figure 3-3 Some of Dalton's symbols.

represented carbon; and a circle with a vertical line represented nitrogen. These cumbersome devices were later abandoned in favor of letters.

The symbols used today consist for the most part of the first letter or the first letter plus another letter from the name if a single letter might lead to confusion with the letter symbol of some other element. For example, the symbols used for hydrogen, carbon, and nitrogen today are H, C, and N, respectively. To avoid confusion with carbon, the symbols for other elements with names beginning with C consist of C plus another letter. Thus, we have Ca for calcium, Cl for chlorine, and Co for cobalt. It should be noted that the first letter of the symbol is capitalized but the second is not. It is important that this convention be strictly adhered to. Considerable confusion results if it is not. For example, if one carelessly wrote CO for cobalt, he would actually be representing a poisonous gaseous compound (carbon monoxide) instead of an element.

In a few cases, the symbol used today comes from the Latin rather than the English name of the element. For example, Na from the Latin word *natrium* is used for sodium, Ag from *argentum* represents silver, and Hg from *hydrargyrum* represents mercury. A complete list of elements with their symbols is found inside the back cover of this text.

The Electron Theory of the Atom

The nuclear atom, as postulated by Rutherford, answers a number of questions. For example, it is easy to understand in retrospect how alpha particles can pass through the gold foil undeflected. However, the nuclear atom also poses a number of questions. We have learned that similarly charged objects repel each other and that oppositely charged bodies attract. What is the nature of the force or forces holding the nucleus together? That is, why does the repulsion of the protons for each other not cause the nucleus to fly apart? What prevents the electrons from being attracted into the nucleus and collapsing the atom?

The first question has plagued physicists and chemists for many years and is still far from being answered. The second question is much closer to being answered. The consideration of this question has led scientists to many truths about the arrangement of the extranuclear electrons. It is this aspect of atomic structure that we consider next. It is very important to the chemist, since it determines the chemical properties of the elements.

The Danish physicist Niels Bohr (1885-1962) proposed a partial answer to the question concerning the forces which prevent the electrons from spiraling into the nucleus. He reasoned that electrons must revolve about the nucleus in fixed orbits, which are associated with a certain energy level. He postulated that as long as electrons remain in these fixed orbits, no energy is lost or gained. That is, the energy of motion counteracts the attractive forces between the nucleus and the electron. Bohr explained the radiant energy emitted by atoms when heated to high temperatures in terms of electrons moving from one orbit to another. That is, when energy of the required intensity is supplied to atoms, electrons jump from the usual energy level (ground state) to a higher energy level. Radiant energy is emitted when the electrons return to their original orbits. The luminosity in the cathode-ray tube is also explained by this phenomenon. In this case the electrons are made to move to higher energy levels by electric energy rather than by heat energy. The luminosity results when electrons of the excited atoms return to their original orbits, causing radiant energy to be emitted.

A practical application of this phenomenon is found in so-called neon lights used for advertising. Neon produces a characteristic orange-red glow when subjected to an electric discharge at low pressures. Again, electrons absorb energy under the influence of electric discharge and move out of their usual energy levels. The radiant energy is emitted when electrons return to their usual energy level. Various colors can be produced in neon signs by using different gases and mixtures of gases. Also the variously colored flames of fireworks are caused by the radiant energy emitted by excited atoms when their electrons return to their normal energy level. Compounds of the chemically reactive metals are the most suitable for this purpose, the characteristic colors produced by the compounds of some of these metals are green, violet, and yellow for compounds of barium, potassium, and sodium, respectively.

Bohr's work was only the beginning in this field. Intensive investigations into the nature of electron arrangement and movement of electrons in atoms (a science known as quantum mechanics) have greatly altered the concept of the atom as presented by Bohr. We no longer think of electrons as revolving about the nucleus in orbits; however, the concept of energy levels is still accepted.

35 THE ELECTRON THEORY OF THE ATOM

Although it is not feasible to present the mathematics and physics behind the modern concept of the atom, it is very helpful to look at a simplified picture of the electron arrangement in atoms given us by quantum mechanics.

ORBITALS

It is not possible for scientists to calculate the exact position of an electron at an exact time; it is possible, however, to define a chunk of space where a particular electron will *probably* be found at a given time. This chunk of space is referred to as an *orbital*.

Orbitals are characterized by three numbers, referred to as *quantum numbers*. The first of these is known as the principal quantum number and is represented by the letter n. Its values correspond to Bohr's energy levels and are related to the sizes of the orbitals. The larger the principal quantum number, the larger the orbital and the further the electrons it contains can get from the nucleus. The principal quantum number is limited to positive integers not including zero (1, 2, 3, 4, . . . , n).

The second quantum number l is related to the shape of the orbital and can have values from 0 up to $n - 1$. That is, if n is 4, l may have values of 0, 1, 2, and 3. The second quantum number corresponds to what is known as the *subenergy level*. That is, if the first two quantum numbers for two orbitals are the same, we say that the electrons in these orbitals are in the same subenergy level. The shapes associated with three values of l are illustrated in Fig. 3-4. We note that when $l = 0$, the shape of the orbital is spherical. When $l = 1$, the orbital consists of two lobes arranged as shown in Fig. 3-4. As the value of l increases, the shapes become considerably more complex.

The third quantum number m is related to the orientation of the orbital in space. Its value ranges from l to $-l$, including zero. If $l = 0$, m can have only the value of 0. In this case, there is only one orientation of the orbital; all orientations of a sphere are equivalent. Furthermore there is only one orbital with the secondary quantum number of zero for each principal quantum number (or in each energy level). If $l = 1$, m may be $+1$, 0, or -1 and there are three possible orientations corresponding to the three directions of the coordinate axis and three orbitals with a secondary quantum number of 1 for each principal quantum number (or in each energy level). When $l = 2$, $m = -2, -1, 0, 1,$ or 2, and there are five orientations and hence five orbitals with the secondary quantum number of 2 for each value of the principal quantum number (or in each energy level).

The relationship of quantum numbers designating the orbitals in the first four energy levels is summarized in Table 3-2. Examination of these data reveals that the number of subenergy levels in an energy level and the number of orbitals in an energy level are n and n^2, respectively. Hence

36 ATOMS

s-orbital ($l = 0$)

p-orbitals ($l = 1$)

d-orbitals ($l = 2$)

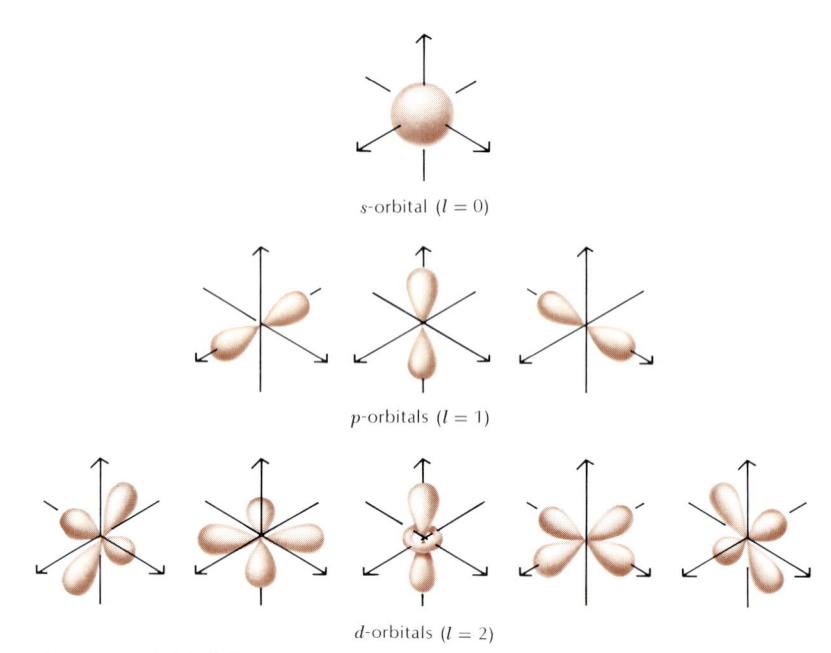

Figure 3-4 Orbital shapes.

the energy level with the principal quantum number of 5 contains 5 subenergy levels and 25 orbitals.

A fourth quantum number, *s*, is used to describe a property of the electrons. This property is roughly analogous to the spin of a sphere about its axis. Since this spin can only be clockwise or counterclockwise, *s* has only two values, $-\frac{1}{2}$ or $+\frac{1}{2}$.

TABLE 3-2 Relation of Quantum Numbers Used to Designate Orbitals

Energy Level n	Subenergy Level l	Orbitals m	Number of Orbitals in Each Subenergy Level	Number of Subenergy Levels in Each Energy Level	Number of Orbitals in Energy Level
1	0	0	1	1	1
2	0	0	1	2	4
	1	−1, 0, +1	3		
	0	0	1		
3	1	−1, 0, +1	3	3	9
	2	−2, −1, 0, +1, +2	5		
	0	0	1		
	1	−1, 0, +1	3		
4	2	−2, −1, 0, +1, +2	5	4	16
	3	−3, −2, −1, 0, +1, +2, +3	7		

THE ELECTRON THEORY OF THE ATOM

ELECTRON CONFIGURATION OF THE ATOMS

The arrangement of electrons in terms of energy levels, subenergy levels, and orbitals is spoken of as the *electron configuration* of the atom. It is now possible to describe the electron configuration of most atoms with a reliable degree of certainty. These electron configurations are governed by the following principles:

1. No two electrons in the same atom can have four identical quantum numbers (this is called the *Pauli exclusion principle* after the Austrian physicist W. Pauli, who enunciated it in 1925). Since s can have only two values, this generalization limits the maximum number of electrons in an orbital to 2.
2. Subenergy levels are usually filled in order of increasing energy. (Note exceptions in principle 4.) This order is defined in terms of n and l. Subenergy levels are filled in the order of $n + l$ value. If more than one subenergy level has the same $n + l$ value, these subenergy levels are filled in the order of increasing value of n. The $n + l$ values for the subenergy levels are shown in Table 3-3. Here the number designates the principal quantum number n, and the secondary quantum number l is designated by the letters s, p, d, and f, which stand for 0, 1, 2, and 3, respectively. The subenergy levels, $5g$, $6f$, $6g$, and $6h$ are hypothetical, since the last electron of element 105 goes into the $6d$ subenergy level. These subenergy levels are indicated to account for the maximum number of electrons possible for these energy levels. This is given by the formula $2n^2$ and is 50 and 72 for the fifth and sixth energy levels respectively. From Table 3-3 we note that the order in which subenergy levels are filled is $1s$, $2s$, $2p$, $3s$, $3p$, $4s$, $3d$, $4p$, $5s$, $4d$, $5p$, $6s$, $4f$, $5d$, $6p$, $7s$, $5f$, $6d$. It is convenient to remember this order; however, should it be forgotten it can easily be derived by the method presented in Table 3-3.
3. Within a subenergy level one electron goes into each orbital before any orbital gets two electrons (*Hund's principle of maximum multiplicity*). This principle is explained by the force of repulsion between negatively charged electrons. That is, if two orbitals of equal energy are accessible one of which contains one electron and the other no electrons, it would take less energy to place the electron in the empty orbital.
4. Completely filled or half-filled subenergy levels acquire extra stability. In the case of energy levels with a secondary quantum number of 2 or 3 (the subenergy levels designated as d and f), this extra stability results in an electron's being transferred from the subenergy level previously filled to the subenergy level being filled if the subenergy level being filled is within one electron of being half-filled or com-

38 ATOMS

TABLE 3-3 $n + l$ Values and Filling Order for Subenergy Levels

Subenergy Level	$n + l$ Value	Order Filled
1s	1 + 0 = 1	1
2s	2 + 0 = 2	2
2p	2 + 1 = 3	3
3s	3 + 0 = 3	4
3p	3 + 1 = 4	5
3d	3 + 2 = 5	7
4s	4 + 0 = 4	6
4p	4 + 1 = 5	8
4d	4 + 2 = 6	10
4f	4 + 3 = 7	13
5s	5 + 0 = 5	9
5p	5 + 1 = 6	11
5d	5 + 2 = 7	14
5f	5 + 3 = 8	17
5g	5 + 4 = 9	X
6s	6 + 0 = 6	12
6p	6 + 1 = 7	15
6d	6 + 2 = 8	X
6f	6 + 3 = 9	X
6g	6 + 4 = 10	X
6h	6 + 5 = 11	X
7s	7 + 0 = 7	16

pletely filled. This, of course, is a deviation from the second generalization.

These generalizations are not without exceptions. However, they are operative in a sufficient number of cases to be considered dependable clues.

There are a number of conventions used in representing electron configurations. The one most commonly encountered uses a number to represent the principal quantum number and the letters s, p, d, and f to represent the secondary quantum numbers 0, 1, 2, and 3 respectively, as in Table 3-3. A superscript is used to indicate the number of electrons in each subenergy level. This is illustrated below:

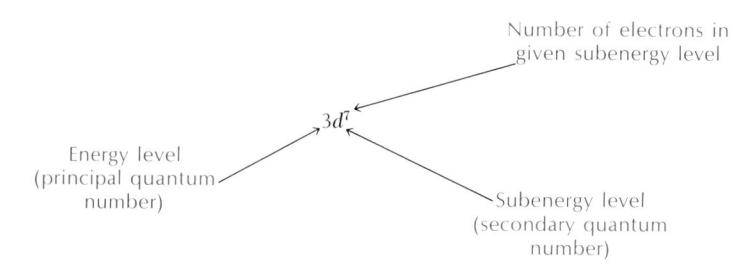

$3d^7$

Energy level (principal quantum number)

Number of electrons in given subenergy level

Subenergy level (secondary quantum number)

39 THE ELECTRON THEORY OF THE ATOM

The subenergy levels are usually listed in order of increasing energy within successive energy levels.

Another method commonly encountered indicates the electron configuration in terms of all four quantum numbers. Here squares are used to represent orbitals, and electrons are represented by arrows. Each group of squares representing a subenergy level is labeled with an appropriate number and letter, as above, to indicate the subenergy level. This is illustrated below:

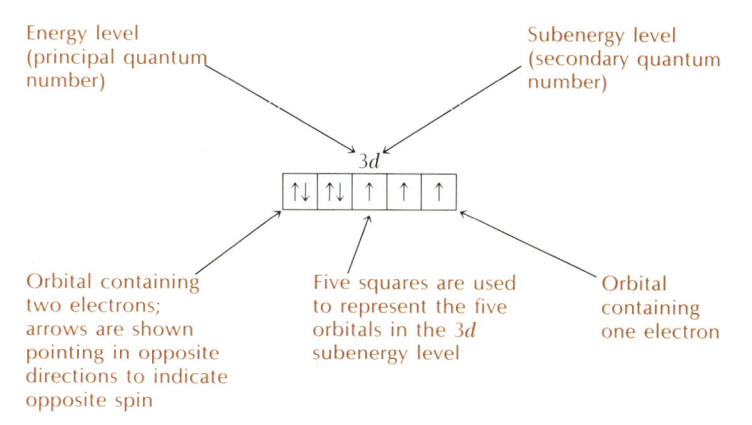

Energy level (principal quantum number) → 3d ← Subenergy level (secondary quantum number)

[↑↓ | ↑↓ | ↑ | ↑ | ↑]

Orbital containing two electrons; arrows are shown pointing in opposite directions to indicate opposite spin

Five squares are used to represent the five orbitals in the 3d subenergy level

Orbital containing one electron

As in the previous method, the subenergy levels are listed in order of increasing energy within successive energy levels.

A third convention simply indicates the number of electrons in each energy level. Here we proceed from the lowest to the highest energy level, and the numbers of electrons for the various energy levels are separated with commas.

These generalizations and conventions become more meaningful with practice. To begin with, hydrogen has an atomic number 1, and hence has one electron in the 1s subenergy level. Representations of the electron configuration of this element by each of the conventions we have discussed are:

$1s^1$

1s
[↑]

1

Since there can be two electrons in the 1s subenergy level, the next element (helium, atomic number = 2) has the following structure:

$1s^2$

1s
[↑↓]

2

Notice that the two arrows are pointing in opposite directions to indicate opposite spins.

With lithium (atomic number = 3) we put the third electron in the 2s subenergy level. Hence we have

$1s^2 \quad 2s^1$

1s 2s
[↑↓] [↑]
2,   1

The 2s energy level is filled at beryllium.

$1s^2 \quad 2s^2$

1s 2s
[↑↓] [↑↓]
2,   2

With boron, we begin adding electrons to the next lowest subenergy level, the 2p.

$1s^2 \quad 2s^2 \quad 2p^1$

1s   2s   2p
[↑↓] [↑↓] [↑][ ][ ]
2,   3

TABLE 3-4 Electron Configurations of Atoms of the Elements Carbon to Neon

Symbol	At. No.	By Subenergy Levels	By Orbital	By Energy Level
C	6	$1s^2, 2s^2, 2p^2$	1s [↑↓] 2s [↑↓] 2p [↑][↑][ ]	2, 4
N	7	$1s^2, 2s^2, 2p^3$	1s [↑↓] 2s [↑↓] 2p [↑][↑][↑]	2, 5
O	8	$1s^2, 2s^2, 2p^4$	1s [↑↓] 2s [↑↓] 2p [↑↓][↑][↑]	2, 6
F	9	$1s^2, 2s^2, 2p^5$	1s [↑↓] 2s [↑↓] 2p [↑↓][↑↓][↑]	2, 7
Ne	10	$1s^2, 2s^2, 2p^6$	1s [↑↓] 2s [↑↓] 2p [↑↓][↑↓][↑↓]	2, 8

41 THE ELECTRON THEORY OF THE ATOM

Note that three squares are labeled $2p$ to represent the three orbitals in the $2p$ subenergy level. Since the $2p$ subenergy level may contain up to six electrons, each of the next five elements [from carbon (atomic number = 6) to neon (atomic number = 10)] adds an additional electron to the $2p$ subenergy level. These electron configurations are indicated in Table 3-4. It should be noted that one electron is placed in each $2p$ orbital before any orbital receives two electrons (see generalization 3 on page 37).

Sodium (atomic number = 11) and magnesium (atomic number = 12) have one and two electrons, respectively, in the $3s$ subenergy level. The next six elements [aluminum (atomic number = 13) to argon (atomic number = 18)] add one electron each to the $3p$ subenergy level, which is filled with six electrons. Note that one electron goes in each orbital in the $3p$ subenergy level before any orbital receives two.

TABLE 3-5 Electron Configurations of the Elements Sodium to Calcium

Symbol	At. No.	By Subenergy Levels	By Orbitals	By Energy Levels
Na	11	$1s^2, 2s^2, 2p^6, 3s^1$		2, 8, 1
Mg	12	$1s^2, 2s^2, 2p^6, 3s^2$		2, 8, 2
Al	13	$1s^2, 2s^2, 2p^6, 3s^2, 3p^1$		2, 8, 3
Si	14	$1s^2, 2s^2, 2p^6, 3s^2, 3p^2$		2, 8, 4
P	15	$1s^2, 2s^2, 2p^6, 3s^2, 3p^3$		2, 8, 5
S	16	$1s^2, 2s^2, 2p^6, 3s^2, 3p^4$		2, 8, 6
Cl	17	$1s^2, 2s^2, 2p^6, 3s^2, 3p^5$		2, 8, 7
Ar	18	$1s^2, 2s^2, 2p^6, 3s^2, 3p^6$		2, 8, 8
K	19	$1s^2, 2s^2, 2p^6, 3s^2, 3p^6, 4s^1$		2, 8, 8, 1
Ca	20	$1s^2, 2s^2, 2p^6, 3s^2, 3p^6, 4s^2$		2, 8, 8, 2

42 ATOMS

Potassium (atomic number = 19) is the first element in which the last electron goes into a higher energy level before the next lowest energy level is filled. This is because the 4s subenergy level is at a lower energy state than the 3d subenergy level. The 4s subenergy level is filled at calcium (atomic number = 20). The electron configurations of elements sodium (atomic number = 11) to calcium (atomic number = 20) are presented in Table 3-5.

With scandium (atomic number = 21) we put the first electron into the 3d subenergy level. The 3d and 4s subenergy levels are filled to their capacities of 10 and 2 electrons, respectively, with zinc (atomic number = 30). The electron configuration of elements with atomic numbers 21 to 30 are shown in Table 3-6. There are two things to be noted about these electron configurations: (1) no orbital in the 3d subenergy level contains two electrons until one electron is put into each orbital; (2) with

TABLE 3-6 Electron Configurations of Elements Scandium to Zinc

(All subenergy levels through 3p are filled. Only the 4s and 3d subenergy levels are shown in this table for electron configurations by subenergy levels and by orbitals. Electron configurations by energy level are complete.)

Symbol	At. No.	By Subenergy Levels	By Orbitals (3d)	(4s)	By Energy Levels
Sc	21	$3d^1, 4s^2$	↑	↑↓	2, 8, 9, 2
Ti	22	$3d^2, 4s^2$	↑ ↑	↑↓	2, 8, 10, 2
V	23	$3d^3, 4s^2$	↑ ↑ ↑	↑↓	2, 8, 11, 2
Cr	24	$3d^5, 4s^1$	↑ ↑ ↑ ↑ ↑	↑	2, 8, 13, 1
Mn	25	$3d^5, 4s^2$	↑ ↑ ↑ ↑ ↑	↑↓	2, 8, 13, 2
Fe	26	$3d^6, 4s^2$	↑↓ ↑ ↑ ↑ ↑	↑↓	2, 8, 14, 2
Co	27	$3d^7, 4s^2$	↑↓ ↑↓ ↑ ↑ ↑	↑↓	2, 8, 15, 2
Ni	28	$3d^8, 4s^2$	↑↓ ↑↓ ↑↓ ↑ ↑	↑↓	2, 8, 16, 2
Cu	29	$3d^{10}, 4s^1$	↑↓ ↑↓ ↑↓ ↑↓ ↑↓	↑	2, 8, 18, 1
Zn	30	$3d^{10}, 4s^2$	↑↓ ↑↓ ↑↓ ↑↓ ↑↓	↑↓	2, 8, 18, 2

elements chromium (atomic number = 24) and copper (atomic number = 29) we encounter for the first time examples of generalization 4. In atoms of both elements, the number of electrons in the 4s subenergy level is only one instead of two. The 4s and 3d subenergy levels are very close together, and half-filled and completely filled subenergy levels acquire extra stability. Thus the chromium atom gains stability by transferring an electron from the 4s subenergy level to the 3d subenergy level to give a half-filled 3d subenergy level. Similarly, the copper atom acquires extra stability by shifting an electron from the 4s to the 3d subenergy level to give a completely filled 3d subenergy level.

This general pattern could be continued until the electron configuration of all the 105 elements had been designated. The student can probably arrive at the electron configuration of most of the remaining elements with the information given. However, some exceptions to the general pattern are encountered. These exceptions are all found in the d and f subenergy levels.

Questions

1 How does the emission of an alpha particle from a radioactive atom affect the atomic number? The mass number?
2 Compare the following with regard to charge and mass:
   a Alpha particles   b Protons
   c Electrons   d Neutrons
   e Beta particles
3 Some of Dalton's postulates regarding the atom have been found incorrect. Identify several.
4 What experimental evidence can you cite to support the following?
   a Atoms of gold are very close to each other.
   b The gold atom is a very porous particle.
   c The mass of the atom is concentrated in a very small part of its total volume.
5 Is it possible for two atoms of different elements to have the same mass number? The same atomic number? Explain your answers.
6 In what ways do atoms of isotopes differ? In what ways are they the same?
7 There are eight isotopes of iron with mass numbers of 52, 53, 54, 55, 56, 57, 58, and 59. How many protons and neutrons are there in each of the eight isotopes?
8 Consider a hypothetical element Y with 28 protons and 26 neutrons in each atom.
   a What is the number of nucleons per atom?
   b What is the mass number?
   c What is the atomic number?
   d How many electrons are there in each atom?

9 Complete the following table:

Symbol	At. No.	Mass No.	No. of Electrons	No. of Protons	No. of Nucleons	No. of Neutrons
P					31	
		209		83		
	82				207	
		115		49		
Na					23	

10 What is the maximum number of electrons that can exist in each of the following?
  a An orbital
  b The first energy level
  c The p subenergy level of a given energy level
  d The s subenergy level of a given energy level
  e The d subenergy level of a given energy level
  f The third energy level
  g The f subenergy level of a given energy level

11 Which of the following sets of numbers represent impossible combinations for quantum numbers? The numbers are arranged in the order of $n, l, m,$ and s.
  a $1, 0, 0, -\frac{1}{2}$ **b** $2, 3, 1, +\frac{1}{2}$
  c $2, 1, 2, -\frac{1}{2}$ **d** $-1, 0, 0, +\frac{1}{2}$
  e $3, 2, -1, +\frac{1}{2}$ **f** $2, -1, 0, -\frac{1}{2}$

12 What are the four quantum numbers for each electron in an atom of scandium?

13 Indicate the electron configurations by each of the three methods described in the chapter for the following elements:
  a Be   **b** Ne   **c** P   **d** Sc
  e Mn   **f** Cu   **g** Ga   **h** Rb
  i Sn   **j** Co   **k** Te   **l** Ra

14 Match the elements in the left-hand column with the electron-configuration endings in the right-hand column:

Magnesium	$3p^1$
Tin	$4s^2$
Aluminum	$3s^2$
Scandium	$4s^1$
Arsenic	$4p^4$
Argon	$5p^2$
Copper	$4p^3$
Selenium	$3p^6$

15 Consider a hypothetical element R with three electrons in the $3d$ subenergy level.
   a What is the number of protons per atom?
   b What is the number of electrons per atom in orbitals with a secondary quantum number of 0?
   c What is the number of orbitals which contain electron pairs in the third energy level?
16 How are atomic weights determined?
17 The atomic weights of isotopes are very near whole numbers. However, the atomic weight for an element may deviate considerably from a whole number. For example, the atomic weights of the naturally occurring isotopes of chlorine are 34.969 and 36.966, but the atomic weight of the element chlorine is 35.453. Explain.

CHAPTER 4 IONS

The world's population explosion has caused fresh water to become a precious resource. The oceans offer a possible supply of this valuable commodity. However, seawater contains large quantities of dissolved ions. The most important of these are Cl^- (1.90 percent), Na^+ (1.06 percent), SO_4^{--} (0.26 percent), Mg^{++} (0.13 percent), Ca^{++} (0.04 percent). In order to make seawater suitable for domestic consumption as well as for many industrial purposes, most of the dissolved ions must be removed. The plant shown in the photograph is designed to accomplish this by distillation; that is, the water is evaporated by heating, leaving behind the nonvolatile ions. The vaporized water is subsequently condensed. (Ray Manley, Stearns-Roger Corporation, Denver, Colo.)

CHAPTER 4 IONS

In the last chapter we learned that there are over 1,400 different kinds of atoms. We also learned that when atoms having the same atomic number are grouped together, there are 105 different groups, or categories, each of which is called an *element*. Because of the intensive effort being made to create new atoms, the number of kinds of atoms and hence the number of elements is likely to increase. All atoms of the same element have identical electron configurations and therefore identical chemical properties. This is true of all isotopes of a given element; chemical properties are determined by electron configurations, and in this respect atoms of the same element do not vary.

The 105 elements combine to form a multitude of compounds. Almost 2 million compounds have been identified and characterized. There seems to be no limit to the number of compounds which can be formed.

The forces bonding one atom to another are called *chemical bonds* (they are a result of the interaction of electrons between atoms). In general, bonds can be classified into two types. One of these types is formed by the transfer of one or more electrons from one atom to another and is called an *ionic* or *electrovalent* bond. The second basic type of bonding results from the sharing of electrons between atoms and is called *covalent* bonding. Although it is common to speak of chemical bonds as being either ionic or covalent, it should be remembered that chemical bonds are rarely completely ionic or covalent but something intermediate. This matter will be discussed in greater detail in the next chapter, along with covalent bonding. Ionic bonding will be covered in this chapter. However, before we begin this discussion of ionic bonding, it is necessary to learn about the periodic classification of the elements.

The Periodic Classification of the Elements

The electron configuration of the atoms presented in the previous chapter is supported by the experimental work of a large number of scientists working in the field of spectroscopy (the science which deals with the study of radiant energy emitted by excited atoms). The scope of this text does not permit us to discuss the work of spectroscopists, but another source of impressive support for the electron arrangements is the periodic table (see Table 4-1).

This table was developed long before scientists were aware of atomic numbers or electrons. The basis for its development was the arrangement of elements according to increasing atomic weights. The result is a table in which the elements are grouped according to such properties as color, melting points of the element and/or its compounds, capacity for chemical combination, and stability of compounds. After the concept of atomic number was developed, it was shown that the periodic table is an arrangement of elements according to increasing atomic number. The basis for the modern periodic table is the periodic law. It states that *the proper-*

TABLE 4-1 The Periodic Table

	IA	IIA		IIIB	IVB	VB	VIB	VIIB		VIII		IB	IIB	IIIA	IVA	VA	VIA	VIIA	0
1s	1 H																		2 He
2s	3 Li	4 Be												5 B	6 C	7 N	8 O	9 F	10 Ne
3s/3p	11 Na	12 Mg												13 Al	14 Si	15 P	16 S	17 Cl	18 Ar
4s/3d/4p	19 K	20 Ca		21 Sc	22 Ti	23 V	24 Cr	25 Mn	26 Fe	27 Co	28 Ni	29 Cu	30 Zn	31 Ga	32 Ge	33 As	34 Se	35 Br	36 Kr
5s/4d/5p	37 Rb	38 Sr		39 Y	40 Zr	41 Nb	42 Mo	43 Tc	44 Ru	45 Rh	46 Pd	47 Ag	48 Cd	49 In	50 Sn	51 Sb	52 Te	53 I	54 Xe
6s/5d/6p	55 Cs	56 Ba		57 La	72 Hf	73 Ta	74 W	75 Re	76 Os	77 Ir	78 Pt	79 Au	80 Hg	81 Ti	82 Pb	83 Bi	84 Po	85 At	86 Rn
7s/6d	87 Fr	88 Ra		89 Ac	104	105													

4f: 58 Ce | 59 Pr | 60 Nd | 61 Pm | 62 Sm | 63 Eu | 64 Gd | 65 Tb | 66 Dy | 67 Ho | 68 Er | 69 Tm | 70 Yb | 71 Lu

5f: 90 Th | 91 Pa | 92 U | 93 Np | 94 Pu | 95 Am | 96 Cm | 97 Bk | 98 Cf | 99 Es | 100 Fm | 101 Md | 102 No | 103 Lw

Metals
Metalloids
Nonmetals
Rare gases

51 THE PERIODIC CLASSIFICATION OF THE ELEMENTS

ties of the elements are periodic functions of the atomic number of the elements. That is, if the elements are arranged in order of increasing atomic number, certain physical and chemical properties regularly appear at definite intervals in the atomic-number progression. The order of filling of subenergy levels as described in the previous chapter is the framework of the periodic table (see Table 4-2).

The horizontal and vertical rows of the periodic table are referred to as *periods* and *groups*, respectively. The first period contains only two elements, corresponding to the capacity of the first subenergy level for two electrons. The second period contains eight elements, which are placed in two categories, one containing two and the other containing six elements. This classification corresponds to the capacity of the 2s and 2p subenergy levels. With the last element of period 2 (neon) the second energy level is complete. The third period is also divided into two categories, one containing two and the other containing six elements. These correspond to the capacities of the 3s and 3p subenergy levels. Since the capacity of the third energy level is 18 and not 8, the third energy level is not filled at the end of the third period. With the fourth period the situation becomes somewhat more complex, since the elements in this period are divided into three categories of two, ten, and six elements. However, this is readily understood, if we remember that after the 3p subenergy level the order of filling is 4s, 3d, and 4p. Hence elements 19 and 20 fill the 4s subenergy level; the next 10 elements fill the 3d subenergy level; and finally the last six elements fill the 4p subenergy level. Period 5 is similar to period 4. The 5s, 4d, and 5p subenergy levels are filled in the fifth period. In periods 6 and 7 the filling of the 4f and 5f subenergy levels increases the number of elements in each period from 18 to 32. The elements involved with the filling of the *f* subenergy levels are generally shown at the bottom of the periodic table to conserve space.

It is convenient to talk about three categories of elements in terms of the position of the "last" electron. The positions of these categories in the periodic table are shown in Table 4-3. One of these consists of the

TABLE 4-2 The Framework of the Periodic Table

	1s	
2s		2p
3s		3p
4s	3d	4p
5s	4d	5p
6s	5d	6p
7s	6d	
	4f	
	5f	

TABLE 4-3 Classification of Elements According to Subenergy Level of Last Electron

s BLOCK REPRESENTATIVE

	IA	IIA
1s	1 H	
2s	3 Li	4 Be
3s	11 Na	12 Mg
4s	19 K	20 Ca
5s	37 Rb	38 Sr
6s	55 Cs	56 Ba
7s	87 Fr	88 Ra

p BLOCK REPRESENTATIVE

	IIIA	IVA	VA	VIA	VIIA	0
						2 He
2p	5 B	6 C	7 N	8 O	9 F	10 Ne
3p	13 Al	14 Si	15 P	16 S	17 Cl	18 Ar
4p	31 Ga	32 Ge	33 As	34 Se	35 Br	36 Kr
5p	49 In	50 Sn	51 Sb	52 Te	53 I	54 Xe
6p	81 Tl	82 Pb	83 Bi	84 Po	85 At	86 Rn

d BLOCK TRANSITION

	IIIB	IVB	VB	VIB	VIIB	VIII			IB	IIB
3d	21 Sc	22 Ti	23 V	24 Cr	25 Mn	26 Fe	27 Co	28 Ni	29 Cu	30 Zn
4d	39 Y	40 Zr	41 Nb	42 Mo	43 Tc	44 Ru	45 Rh	46 Pd	47 Ag	48 Cd
5d	57 La	72 Hf	73 Ta	74 W	75 Re	76 Os	77 Ir	78 Pt	79 Au	80 Hg
6d	89 Ac	104	105							

f BLOCK RARE EARTHS

4f	58 Ce	59 Pr	60 Nd	61 Pm	62 Sm	63 Eu	64 Gd	65 Tb	66 Dy	67 Ho	68 Er	69 Tm	70 Yb	71 Lu
5f	90 Th	91 Pa	92 U	93 Np	94 Pu	95 Am	96 Cm	97 Bk	98 Cf	99 Es	100 Fm	101 Md	102 No	103 Lw

elements in which the last electron goes into the *s* or *p* subenergy level of the outermost energy level. This category is often referred to as the *representative group*. The representative elements are found in the first two and last six groups (0 and A groups) of the periodic table. The number of the A groups indicates the number of electrons in the outermost energy level of the atoms of each element. Since the number of electrons in the outermost energy level of the atoms determines to a large degree the chemical properties of an element, the elements within a given group are similar in chemical properties.

The *transition group* consists of the center portion of the periodic table (B groups and group VIII). These elements are characterized by the last electron's going into the *d* subenergy level of the next-to-outermost energy level of its atoms. The two outermost energy levels of the atoms of these elements are incomplete. Only in the case of groups IB and IIB does the group number indicate the number of electrons in the outermost energy level. The other B groups and group VIII do not show this property.

Finally, the elements in which the last electron goes into an *f* subenergy level in the third-to-the-outermost energy level of the atoms constitute the *rare earths*, or *inner transition* category. These elements are found in the two horizontal rows at the bottom of the periodic table.

Location in the periodic table also classifies an element as a rare gas, metal, nonmetal, or metalloid. (See Table 4-1.)

All the rare gases are found in group 0. Except for helium, they have eight electrons in the outermost energy level. Helium has only two in the outermost energy level, since this is the maximum capacity for the first energy level. Rare gases are of special significance, since they are considerably less active chemically than the other elements. Because of this lack of chemical activity, the rare gases are said to have a stable electron configuration.

Metals are generally shiny, malleable, good conductors of electricity, and hard solids. In terms of electron configuration they usually have three or fewer electrons in the outermost energy level. Tin and lead, which have four electrons in the outermost energy level, along with bismuth, which has five electrons in the outermost energy level, are exceptions to this generalization.

Except for hydrogen, helium, and beryllium, the *s*, *d*, and *f* blocks consist entirely of metals. The elements in the lower lefthand corner of the *p* block are also metals.

Nonmetals, which are usually solids or gases, are poor conductors; the solids are commonly dull and brittle. One nonmetal, bromine, is a liquid. They are characterized by four or more electrons in the outermost energy level. Hydrogen and boron, which have one and three electrons respectively in the outermost energy levels, are exceptions to this generalization. Except for the 0 group, the elements in the upper righthand corner of the *p* block are nonmetals.

The elements adjacent to the zigzag line in the p block, along with beryllium, constitute the metalloids. They have properties between those of the metals and nonmetals.

Probably the most important feature of the periodic table is that it brings together in a single group elements of similar chemical properties. This similarity is more pronounced in some groups than in others. Where the similarity is particularly clear-cut, the group is referred to as a *family*, and the group is referred to collectively by a family name. We have already mentioned the rare gases, the family name for group 0. Other families are the alkali metals (group IA), the alkaline earth metals (group IIA), the chalcogens (group VIA), and the halogens (group VIIA).

Ionic Bonding   The transfer of electrons from one atom to another results in the formation of charged particles called *ions*. Atoms which accept electrons become negatively charged ions, and atoms which donate electrons become positively charged ions. Ions have different electron configurations than the atoms from which they are derived and hence are entirely different substances. Thus sodium chloride has a set of chemical properties different from those of its constituent elements. Positively and negatively charged ions occur together in groups, frequently pairs. They are held together by electrostatic forces to form a definite geometric pattern. The positive ions are surrounded by negative ions, and negative ions are surrounded by positive ions. These aggregates of ions, called *ionic crystals*, will be discussed in greater detail in a later chapter.

The definition of ionic bonding naturally raises certain questions. What causes atoms to interact to form ions? Which atoms donate electrons? How many electrons are transferred? What conditions favor ion formation?

Let us consider the first question. Atoms interact to acquire a more stable electron configuration. Atoms will not form ions unless this increased stability is acquired. A stable electron configuration may be of several types. We learned that elements in group 0 of the periodic table (rare gases) form only a very limited number of compounds and are much less reactive than other elements. Hence the electron configuration of atoms of any of the rare gases is regarded as a very stable one. Many interactions between atoms result in ions with an electron configuration of one of the rare gases. For representative elements this electron activity involves only the electrons in the outermost energy level. These are called *valence electrons*. In discussing compound formation, it is convenient to use electron formulas, which consist of the symbol of the element with the appropriate number of dots around it to represent the valence electrons. If the electron formula represents an ion, the charge must also be indicated. The electron formulas of isolated atoms of the elements of the first three periods of the periodic table follow:

IONIC BONDING

H·   He:

Li·   ·Be·   ·B̈·   ·C̈·   ·N̈·   ·Ö·   :F̈:   :N̈e:

Na·   ·Mg·   ·Al·   ·Si·   ·P̈·   ·S̈·   :C̈l: :Är:

In forming ions, atoms can acquire a stable electron configuration by transferring electrons out of or accepting electrons into the outermost energy level. For example, chlorine could acquire the electron configuration of neon by losing seven electrons, or it could gain one electron and acquire the electron configuration of argon. Similarly, sodium could gain seven or lose one. Atoms usually do whichever involves the least number of electrons. Thus sodium atoms and atoms of other elements in group IA of the periodic table lose one electron, and chlorine and other halogens gain one. Compound formation between these groups of elements is illustrated below:

$$Na + :\ddot{C}l: \quad [Na^+ + :\ddot{C}l:^-]$$
Atoms   **Ions**

Note that no dots are assigned to the sodium ion, since the outermost energy level is now empty. Eight dots are placed around the chloride ion to represent eight electrons in the outermost energy level.

If we consider compound formation between potassium and sulfur, we note that each potassium atom loses only one and each sulfur atom gains two; hence two potassium atoms are required to react with one sulfur atom:

$$\begin{matrix}K\circ \\ \quad + :\ddot{S}: \\ K\circ\end{matrix} \quad [2K^+ + :\ddot{S}:^{--}]$$

Compound formation between calcium and bromine requires two atoms of bromine for every calcium atom:

$$Ca + \begin{matrix}:\ddot{B}r: \\ :\ddot{B}r:\end{matrix} \quad [Ca^{++} + 2:\ddot{B}r:^-]$$

Each magnesium atom loses two electrons while each nitrogen atom gains three electrons. Therefore compound formation between these two elements requires three magnesium atoms, which lose a total of six electrons, to two nitrogen atoms, which gain a total of six electrons:

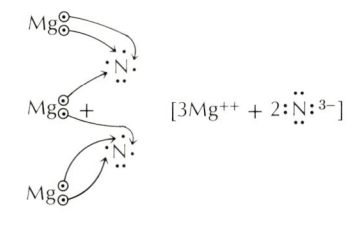

$$\begin{matrix}Mg\circ\!\circ \\ \\ Mg\circ\!\circ + \\ \\ Mg\circ\!\circ\end{matrix} \begin{matrix}:\ddot{N}: \\ \\ :\ddot{N}:\end{matrix} \quad [3Mg^{++} + 2:\ddot{N}:^{3-}]$$

Often it is convenient to represent a compound by its empirical (or simplest) formula rather than by the more cumbersome electron formula. The empirical formula consists simply of the symbols of the combined elements with subscripts to indicate the simplest whole-number ratio of the atoms in the compound. The empirical formulas for the compounds previously discussed are NaCl, K_2S, $CaBr_2$, and Mg_3N_2.

Since metals generally have three or fewer valence electrons and nonmetals generally have four or more, atoms of metals usually lose electrons and atoms of nonmetals usually gain electrons when ionic bonds are formed. However, there is a difference in the ease with which metals lose electrons and the ease with which nonmetals acquire them. Since electrons are held in their orbitals by electrostatic attraction, energy is required to remove them. This energy, which can be measured by experimentation, is called *ionization energy* if measured in calories and *ionization potential* if measured in volts. The energy required to remove the first electron is called the first ionization energy (Table 4-4), and the energy required to remove the second electron is called the second ionization energy, etc. Each successive ionization energy is larger than the one preceding it. The fourth ionization energy is generally so large that the fourth electron is rarely removed in ion formation. Ions with a charge of +4 are unusual.

Among the factors which influence the size of the ionization energies are the distances between the electron to be removed and the nucleus and the nuclear charges.

The larger the atom, the greater the distance between the negative electron and the positive nucleus and hence the lower the ionization energy. The larger the nuclear charge, the greater the attraction for the outer electrons and the larger the ionization energy. However, this factor is complicated by the screening effect of the electrons separating the nucleus from the valence electrons.

In general, the size of the atom and nuclear charge increase with atomic number within a group of the periodic table. These factors have an opposing effect upon each other. However, the former seems to be the more important factor, since the ionization energy usually decreases as one goes down the group. Some exceptions to this generalization are found among the transition elements.

As one moves across the periodic table, the size of the atom decreases and the nuclear charge increases. Both factors lead to an increase in the ionization energy. Since metals react by losing electrons, the chemical activity of metals increases as one moves down within the group and decreases as one moves from the left to the center of the periodic table within a period.

Electron affinity is an indication of the ease with which an atom accepts an electron. It is defined as the energy involved when a single electron

TABLE 4-4 Ionization Energies of the Elements in Kilocalories per Gram Atomic Weight

	IA	IIA
1s	H 313	
2s	Li 124	Be 215
3s	Na 118	Mg 176
4s	K 100	Ca 141
5s	Rb 96	Sr 131
6s	Cs 90	Ba 120
7s	Fr	Ra 122

	IIIB	IVB	VB	VIB	VIIB		VIII		IB	IIB
3d	Sc 154	Ti 158	V 155	Cr 156	Mn 171	Fe 182	Co 181	Ni 176	Cu 178	Zn 216
4d	Y 141	Zr 160	Nb 157	Mo 164	Tc	Ru 169	Rh 172	Pd 192	Ag 175	Cd 207
5d	La 129	Hf	Ta	W 184	Re 181	Os 201	Ir 212	Pt 207	Au 213	Hg 240
6d	Ac	(104)	(105)							

	IIIA	IVA	VA	VIA	VIIA	0
1s						He 567
2p	B 191	C 260	N 325	O 314	F 402	Ne 495
3p	Al 138	Si 188	P 243	S 239	Cl 300	Ar 363
4p	Ga 138	Ge 181	As 226	Se 225	Br 273	Kr 322
5p	In 133	Sn 169	Sb 199	Te 208	I 241	Xe 279
6p	Tl 141	Pb 171	Bi 195	Po 194	At	Rn 248

4f	Ce 159	Pr 132	Nd 145	Pm	Sm 129	Eu 131	Gd 142	Tb 155	Dy 157	Ho	Er	Tm	Yb 143	Lu 142
5f	Th	Pa	U	Np	Pu	Am	Cm	Bk	Cf	Es	Fm	Md	No	Lw

is added to a single atom. A high electron affinity indicates that the atom will accept an electron readily to form a negative ion. Small atoms and high nuclear charges favor high electron affinities and ion formation. Hence, the chemical activity of nonmetals decreases as one goes down a group and decreases as one moves from the right to the center of the periodic table. Trends in the periodic table are summarized in Table 4-5.

We have spoken of ion formation only in terms of acquiring electron configurations of one of the rare gases. Simple monatomic negative ions are generally of this type. Also the positive ions formed from atoms of the metals of groups IA, IIA, and IIIB and aluminum have electron configurations of one of the rare gases (Table 4-6). However, there are other stable electron configurations for atoms of metals. One of these is an outer shell of 18 electrons, characteristic of the metals of groups IB and IIB. The heavier elements in groups IVA and VA [tin (50), lead (82), antimony (51), and bismuth (83)] acquire stable electron configurations consisting of two electrons beyond a shell of 18 by losing the p valence electrons but keeping the s valence electrons. The transition metals between groups IIIB and IIB form stable electron configurations with one to nine electrons in d orbitals beyond a rare-gas electron configuration. Most of the metals in this group form more than one ion with different charges. These ions are formed by losing the electrons in the outermost s subenergy level and zero, one, or two electrons from the outermost d subenergy level. The most common charges for ions from this category are $+2$ or $+3$. We have learned previously that atoms acquire extra stability with a half-filled or filled d subenergy level. This is manifested in the formation of ions by transition metals. For example, the manganese atom, which has five electrons in the $3d$ subenergy level (one in each orbital) and a filled $4s$ subenergy level, forms only one positive ion, Mn^{++}.

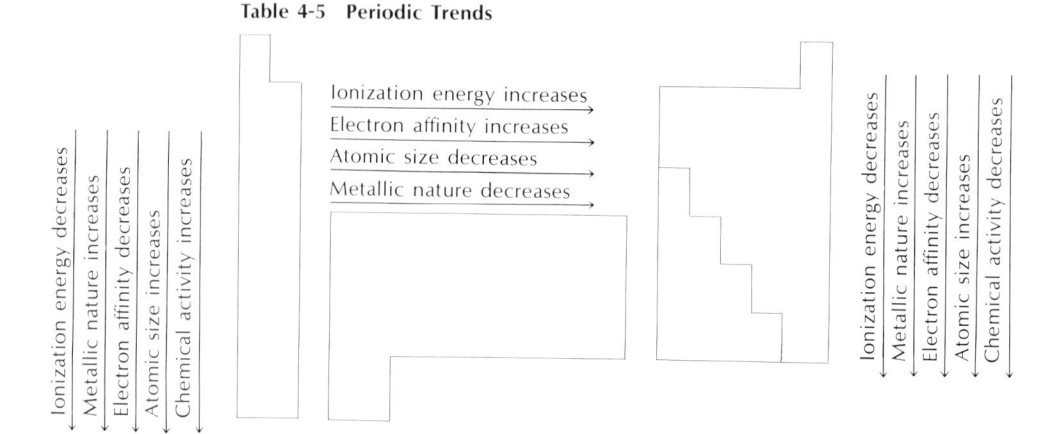

Table 4-5 Periodic Trends

TABLE 4-6 Types of Monatomic Ions

	IA	IIA
1s	1 H	
2s	3 Li	4 Be
3s	11 Na	12 Mg
4s	19 K	20 Ca
5s	37 Rb	38 Sr
6s	55 Cs	56 Ba
7s	87 Fr	88 Ra

	IIIB	IVB	VB	VIB	VIIB	VIII			IB	IIB
3d	21 Sc	22 Ti	23 V	24 Cr	25 Mn	26 Fe	27 Co	28 Ni	29 Cu	30 Zn
4d	39 Y	40 Zr	41 Nb	42 Mo	43 Tc	44 Ru	45 Rh	46 Pd	47 Ag	48 Cd
5d	57 La	72 Hf	73 Ta	74 W	75 Re	76 Os	77 Ir	78 Pt	79 Au	80 Hg
6d	89 Ac	104	105

	IIIA	IVA	VA	VIA	VIIA	0
						2 He
2p	5 B	6 C	7 N	8 O	9 F	10 Ne
3p	13 Al	14 Si	15 P	16 S	17 Cl	18 Ar
4p	31 Ga	32 Ge	33 As	34 Se	35 Br	36 Kr
5p	49 In	50 Sn	51 Sb	52 Te	53 I	54 Xe
6p	81 Ti	82 Pb	83 Bi	84 Po	85 At	86 Rn

Negative ions with electron configuration of one of the rare gases
Positive ions with electron configuration of one of the rare gases
Positive ions with stable electron configurations of one to nine electrons in d orbitals beyond a rare-gas configuration
Positive ions with stable electron configurations of 18 electrons in the outermost energy level
Positive ions with stable electron configurations of two electrons beyond an energy level containing eight electrons

60 IONS

It never loses d electrons. Iron, which has six electrons in the $3d$ subenergy level (four orbitals with one electron in each and one orbital containing two electrons) forms two ions. One has a charge of $+2$ and is formed by the loss of the $4s$ electrons. The other has a charge of $+3$ and is formed by the loss of the $4s$ electrons and one electron from the $3d$ orbital containing two electrons, resulting in a half-filled $3d$ orbital. The latter ion is more common and more stable.

Questions

1 What is the essential difference between an ionic bond and a covalent bond?
2 How is an atom changed by the formation of an ionic bond? What portion of the atom remains unchanged?
3 How does an ion differ from an atom?
4 What is the mechanism by which ionic compounds are formed? What is the nature of the force that holds ions together in a crystal?
5 In terms of atomic structure, what similarities do elements of groups IA and IB possess?
6 Select from each pair of elements the one which you think has each of the following characteristics:
  a Exhibits larger ionization energy (1) Br, I
  b Is more metallic (2) S, Cl
  c Loses its valence electrons more (3) Rb, Cs
    readily (4) Mg, Na
  d Is more reactive chemically (5) Sn, Pb
  e Has smaller atom (6) C, N
7 Indicate how the following vary as one proceeds (1) left to right in a period and (2) top to bottom in a group:
  a Size of atoms
  b Ionization energy
  c Electron affinity
  d Chemical activity
  e Metallic characteristics
  f Nonmetallic characteristics
8 If an element with the atomic number of 18 is a rare gas, in what periodic groups would you expect to find elements with atomic numbers of 17 and 19?
9 Explain or account for the following:
  a As one proceeds down group IVA, the elements become more metallic and less nonmetallic.
  b Sodium forms only ions with a $+1$ charge but calcium forms only ions with $+2$ charge.
  c There are 18 elements in the fourth period of the periodic table but only 8 elements in the third period of the periodic table.

61 QUESTIONS

10 How many valence electrons do atoms of the following elements have?
  a Hydrogen   b Sulfur   c Lithium

11 Complete the following table:

At. No.	No. of Valence Electrons	Period	Group	Representative, Transition, or Rare Earth	Metal, Nonmetal, or Metalloid	Number of Energy Levels
20						
6						
17						
39						
55						
48						
53						

12 Without consulting the periodic table, indicate the groups in which the elements having the following atomic numbers would be found.
  a 3   b 6   c 10
  d 12   e 17   f 21
  g 25   h 34

13 Why is the second ionization energy always larger than the first?

14 Predict which element will have the lowest first ionization energy and which element will have the highest electron affinity.

15 Is there a limit to the number of electrons that an atom will lose to form an ion? Explain your answer.

16 What are the electron formulas for the following?
  a Rubidium   b Silicon   c Barium
  d Iodine   e Hydrogen   f Radon
  g Selenium   h Gallium   i Antimony

17 Consider the subenergy levels designated as s, p, d, and f.
  a In what energy level does each of the subenergy levels first appear?
  b What is the maximum number of orbitals in each subenergy level?
  c Which subenergy level contains the last electron of an atom of a rare-earth element?
  d In which subenergy level do the orbitals have only one possible orientation in space?
  e Which subenergy level contains the last electron of a halogen?

62 IONS

f Which subenergy level contains the last electron of an alkali metal?

18 Each atom of element X has two valence electrons and each atom of Y has five valence electrons. Predict which element:
   a Is the better electrical conductor
   b Has the higher ionization energy
   c Is more likely to be a gas
   d Is more likely to be a shiny solid
   e Has the higher electron affinity
   Write the electron formula for the compound between X and Y.

19 By using the principles learned about the periodic table, write the formulas of the missing compounds:

NaCl	KF	XXX	MgO
KCl	KCl		CaO
	KBr	YBr_3	SrO
CsCl	KI	$LaBr_3$	
FrCl		$AcBr_3$	RaO

20 Which of the following formulas represent stable ionic compounds?
   a FrI   **b** $RbAt_3$ **c** $PbCl_2$
   d Ra_2As **e** KH   **f** CaSe
   g BiF_3

21 Elements X and Y combine to form $[2X^+ + :\ddot{Y}:^{--}]$. What does this tell you about the structures of the atoms of X and Y?

22 List four ions that have the same electron configurations as each of the rare gases listed below:
   a Ne   **b** Ar   **c** Kr

23 Write the electron formulas for compounds formed from the reactions between the following pairs of elements:
   a Rb, F   **b** Sr, S   **c** Ca, P
   d Ra, I   **e** Cs, O   **f** K, N

24 In the formation of ions, atoms acquire one of four main types of stable electron configurations. What are these four types?

25 Classify the following ions according to the number of electrons in the outermost energy level:
   a Ba^{++} **b** Cu^{++} **c** Mn^{++}
   d Fe^{3+} **e** Cl^- **f** H^-
   g Sn^{++} **h** K^+ **i** Sc^{3+}
   j Ag^+ **k** S^{--} **l** Bi^{3+}

26 In what group and what period would one find each of the elements whose electron configurations end as shown?

   a   2s   **b**   3p
        [↑]   [↑↓ | ↑ | ↑]

c   5s   4d
    [↑↓]   [↑][↑][↑][↑][↑]

d   2p
    [↑↓][↑↓][↑↓]

e   4s   3d
    [↑↓]   [↑][ ][ ][ ][ ]

27 Which of the following statements are *not* true about an atom of the element with atomic number of 24?
   a It is a transition element.
   b It has five electrons in the third energy level.
   c The first two quantum numbers of the last electron are $n = 3$ and $l = 3$.
   d It is listed in group VIB of the periodic table.
   e It is listed in the third period of the periodic table.
   f It has 25 protons in the nucleus of each atom.
   g It is a nonmetal.

28 By consulting the periodic table, identify each of the following:
   a An element which has five electrons in each atom
   b An element with three unpaired electrons in each atom
   c An element with a half-filled d subenergy level in each atom
   d An alkali metal
   e An element with seven electrons in the outermost energy level of each atom
   f A rare gas
   g An element whose atoms have a completely filled $3d$ subenergy level and only one electron in the $4s$ subenergy level
   h An element which forms $+2$ ions
   i An element which forms -3 ions
   j A transition element
   k An element which forms ions by losing only p electrons
   l An element which forms ions by losing s and d electrons
   m A transition element which forms ions by losing only s electrons
   n A representative element which forms ions by losing only s electrons
   o An element whose atoms have two filled $3d$ orbitals
   p An element whose atoms have one filled $4p$ orbital
   q A rare earth
   r A halogen
   s An element whose atoms have a half-filled second energy level
   t An element which forms a negative ion with an electron configuration which is the same as that of helium
   u An element whose atoms have electron configurations which end with s^2p^2
   v An element with only one electron in the $3p$ subenergy level

w An element which has three electrons in the 3d subenergy level
x An element which forms very few, if any, compounds

29 In some periodic tables hydrogen appears as the first element in group VIIA. Can you offer any justification for this?

30 Consider element X in period 3 and group VIA of the periodic table.
  a What is the atomic number of X?
  b What are the first two quantum numbers, n and l, of the last electron?
  c Is the last electron paired or unpaired?
  d If element X reacted with element W in period 2 and group IIA of the periodic table, what would be the electron formula of the compound?

31 Consider the following information about the hypothetical elements milwaukium and wisconsine:

	Milwaukium	Wisconsine
Symbol	M	W
Atomic number	20	17
Density	1.31 g/ml	2.10 g/liter

  a Write the electron configuration of M by each of the three methods you have learned.
  b In what period of the periodic table would M be found?
  c In what group of the periodic table would W be found?
  d Write the electron formula for a possible compound of M and W.
  e Classify W as a metal, nonmetal, or rare gas.
  f Classify M as a representative, transition, or rare-earth element.
  g What is the weight of 312 ml of milwaukium?
  h What is the volume of 1.00 lb of wisconsine?

32 Consider the following hypothetical elements with the indicated number of electrons in the given subenergy levels:

Symbol	Electrons	
R	$3d^3$	
S	$2p^5$	
T	$4s^2$	(none in 3d)
U	$3p^6$	(none in 4s)
V	$2p^4$	
W	$2s^2$	(none in 2p)
X	$3p^3$	
Y	$3s^1$	
Z	$3p^2$	

a Which are rare gases?
b Which are transition elements?
c Which are representative metals?
d What is the atomic number of W?
e How many protons are there in each atom of Z?
f How many electrons with a secondary quantum number of 0 are there in each atom of R?
g In what group of the periodic table is Y?
h In what period of the periodic table is U?
i What are the principle and secondary quantum numbers of the last electron in V? $n =$; $l =$
j How many electrons are there in the outermost energy level of Z?
k Using the hypothetical symbols from page 64, write the electron formula for the compound between X and T.
l Using the hypothetical symbols from page 64, write the electron formula for the compound between V and Y.

33 Copy the following diagram which represents the first three periods of the periodic table. Place the symbols for the hypothetical elements described below in the proper places in your diagram. Upon completion the table should have no empty spaces, and all hypothetical elements listed below should be in the table. Place the symbols on the table consecutively, that is, in the order in which the hypothetical elements are described below.

a X is the most active metal in this portion of the table.
b Y is the most active nonmetal in the table.
c W has only one proton in the nucleus.
d R and T have eight electrons in the outermost energy level. The T atom is larger than the R atom.
e Atoms of the elements O and P contain three unpaired electrons. P is more metallic than O.
f Atoms of Z contain seven electrons in the outermost energy level.
g Ki forms +2 ions with electron structures identical to atoms of R.
h Hr forms −2 ions with the same electron structure as atoms of T.
i Element J has one electron in the outermost energy level.

j Elements No and So form the compounds SoZ_4 and NoZ_4. No is more electronegative than So.
k Pr is a rare gas.
l The first two quantum numbers for the last electron in the Bs atom are $n = 2$ and $l = 0$.
m Elements Go and Ld are in period 2. Ld has a higher ionization energy than Go.
n Element Tx is a metalloid.

34 The first periodic tables were written with the elements arranged in order of increasing atomic weights. The modern periodic tables are written with the elements arranged in order of increasing atomic number. Locate two places on the periodic table where elements would be placed in the wrong group if arranged in order of increasing atomic weight.

CHAPTER 5 DISCRETE MOLECULES AND POLYATOMIC IONS

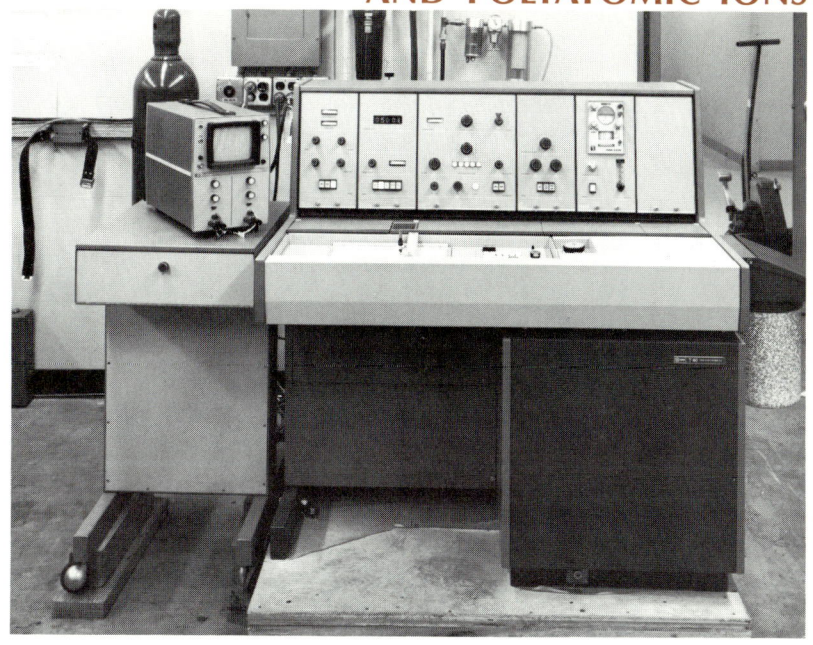

Verifications of proposed geometries for molecules are frequently provided by instruments such as the *nuclear magnetic resonance spectrometer*. This instrument provides information which permits calculation of bond angles and the number of hydrogens bonded to each carbon atom. (Varian Associates, Palo Alto, Calif.)

CHAPTER 5   DISCRETE MOLECULES AND POLYATOMIC IONS

The ionic bond is formed between atoms of very dissimilar elements. In fact, it is this very dissimilarity which is responsible for the formation of the bond. The metal atom, losing electrons to attain a stable electron configuration, becomes a positive ion, and the nonmetal atom becomes a negative ion by gaining electrons. The bond is the result of the electrostatic attraction between the positive and the negative particles.

Chemical bonds are formed not only between dissimilar atoms but between similar or even identical atoms. The common gaseous elements, hydrogen, nitrogen, oxygen, and chlorine, each have as structural unit a *molecule*, which is composed of two atoms of the element bonded together. That the nature of this bond is quite different from that of the ionic bond should be obvious. In the chlorine molecule, for instance, neither atom has a greater tendency to lose or to gain an electron than the other. However, each atom has a strong tendency to attain the stable configuration of a rare gas. This configuration can be attained by both atoms by sharing a pair of electrons, each atom contributing one electron of the pair. The electron formula for the chlorine molecule is

$$:\!\overset{..}{\underset{..}{Cl}}\!:\!\overset{..}{\underset{..}{Cl}}\!:$$

The shared pair of electrons is attracted equally to the two nuclei, or—stated another way—the shared pair of electrons spends as much time in the vicinity of one nucleus as in the vicinity of the other. This shared pair of electrons constitutes what is called the *covalent bond*.

The hydrogen molecule has the electron formula

$$H:H$$

since the hydrogen atom attains the electron configuration of the rare gas helium by gaining one electron.

The electrostatic force of the ionic bond is exerted equally in all directions. A positive ion attracts a negative ion above it with exactly the same force that it attracts a similar ion at the same distance below it. Covalent bonds are completely different in this respect. They are the result of the sharing of a pair of electrons. The electrons of the bond occupy a specific orbital which belongs to both bonded atoms. This *bond orbital* can be viewed as the result of the overlapping of two atomic orbitals. Thus the electron pair of the covalent bond is restricted to a specific portion of space, and the attraction between the two atoms which is manifested in this covalent bond is directional. The electrostatic force, although effective over relatively short distances, is not limited and extends out with decreasing intensity to infinite distances. Again, the covalent bond differs. There is no bond until overlap of the two orbitals is brought about. Thus covalent bonds are limited in extent. These short-range, directional bonds result in the discrete structural unit called the *molecule*, which is composed of atoms of one or more elements.

The discussion of molecular compounds requires a third type of chemical formula. We have already made considerable use of electron formulas and empirical, or simplest, formulas. The third type, the *molecular formula*, has meaning only when applied to a substance in which the structural unit is a molecule. The molecular formula indicates by subscripts the number of atoms of each element present in the molecule. Ionic substances such as NaCl, K_2S, $CaBr_2$, and Mg_3N_2, discussed in Chap. 4, consist of positive and negative ions combined in a specific ratio but not combined into discrete molecules. Thus these substances can be represented *only* by empirical or electron formulas. Molecular substances can be represented by any of the three types of formulas. Molecular formulas for chlorine and hydrogen are Cl_2 and H_2, respectively. They tell the exact constitution of each molecule. The chlorine molecule consists of two chlorine atoms, and the hydrogen molecule consists of two hydrogen atoms. Electron formulas for the two molecules are $:\overset{..}{\underset{..}{Cl}}:\overset{..}{\underset{..}{Cl}}:$ and H:H. Empirical, or simplest, formulas for the two substances would be simply Cl and H.

We must be careful in our consideration of the covalent bond not to oversimplify the situation. The chlorine atom has one half-filled $3p$ orbital which must be filled to give the atom a stable octet. However, once the chlorine molecule has formed, it is not reasonable to consider that the electron pair of the bond occupies a portion of space which is simply the result of the overlapping of the $3p$ atomic orbitals. With the two chlorine nuclei and the total of 34 electrons now held in close proximity and under each other's influence, a mutual rearrangement of the extranuclear electrons takes place. None of the electrons can be considered to occupy their prebonding orbitals. Although the molecular orbitals are different from the atomic orbitals, important information about their nature will be obtained from a consideration of the atomic orbitals from which they were formed. We shall discuss this relationship in greater detail subsequently.

The Polar Covalent Bond

In addition to molecules formed between two identical atoms, molecules are formed between similar atoms. An atom of hydrogen has a greater tendency to lose an electron than an atom of chlorine does, and the chlorine atom has a greater tendency to gain an electron than the atom of hydrogen does, but these differences are not great enough to bring about a complete transfer of an electron. The two atoms, instead, share an electron pair to form a molecule of hydrogen chloride:

$$H:\overset{..}{\underset{..}{Cl}}:$$

THE POLAR COVALENT BOND

But another factor must now be considered. We can no longer say that the electron pair is equally attracted to the two nuclei, for the larger, more highly charged chlorine nucleus has a stronger attraction for electrons than does the small, singly charged hydrogen nucleus. As a result the molecule is not electrically symmetrical as the molecules of hydrogen and of chlorine are. The fact that there is a greater probability of finding the electron pair in the vicinity of the chlorine nucleus than in the vicinity of the hydrogen nucleus results in the chlorine "end" of the molecule being slightly more negative than the hydrogen "end." We say that the molecule is *polar* or that it is an electric dipole. The bond in this case is called a *polar covalent bond*, while that of the hydrogen or chlorine molecules is a *nonpolar covalent bond*.

In our search for an explanation for the properties of matter it has become apparent that the explanation for the formation of a chemical bond between atoms begins with the tendency for atoms to attain a stable electron configuration. The fact that for many atoms this stable configuration requires eight electrons in the outer energy level is referred to as the *octet rule*.

The nature of the bond between atoms is determined in large measure by the method by which the octet rule is satisfied—whether by gaining, losing, or sharing electrons. As a next step we wish to be able to *predict* what kind of a bond will form between any two atoms. Among the active metals of groups IA and IIA we have already observed that the tendency to lose valence electrons increases with increasing atomic size. The *ionization energy*, or energy necessary to remove a valence electron, is a measure of this property. Among the active nonmetals of groups VIA and VIIA, the tendency to gain electrons increases with decreasing atomic size. The *electron affinity* is a measure of this property. Neither of these properties tells us all we need to know, however, since neither can be used satisfactorily to describe the properties of the atoms of the large number of elements in the middle of the periodic table.

The need is for a single index to indicate the attraction of an atom for the electrons of a bond—and this has been devised. It is called the *electronegativity*. On the electronegativity scale an atom of fluorine, which attracts electrons more strongly than any other atom, is assigned the maximum value of 4.0. Other elements are assigned values relative to this through a consideration of their electron affinities and ionization energies (see Table 5-1).

As we consider one atom after another in the Table 5-1, we note the gradual change in strength with which an atom attracts electrons. We have described the bond which forms between an atom of sodium and one of chlorine with respective electronegativities of 0.9 and 3.0 as an ionic bond and that which forms between two identical atoms as a covalent bond. How shall we describe or name the thousands of bond

TABLE 5-1 The Electronegativities of the Elements

	IA	IIA	IIIB	IVB	VB	VIB	VIIB	VIII			IB	IIB	IIIA	IVA	VA	VIA	VIIA	0
1s	1 **H** 2.1																	2 **He**
2s	3 **Li** 1.0	4 **Be** 1.5																
2p													5 **B** 2.0	6 **C** 2.5	7 **N** 3.0	8 **O** 3.5	9 **F** 4.0	10 **Ne**
3s	11 **Na** 0.9	12 **Mg** 1.2																
3d			21 **Sc** 1.3	22 **Ti** 1.5	23 **V** 1.6	24 **Cr** 1.6	25 **Mn** 1.5	26 **Fe** 1.8	27 **Co** 1.8	28 **Ni** 1.8	29 **Cu** 1.9	30 **Zn** 1.6						
3p													13 **Al** 1.5	14 **Si** 1.8	15 **P** 2.1	16 **S** 2.5	17 **Cl** 3.0	18 **Ar**
4s	19 **K** 0.8	20 **Ca** 1.0																
4d			39 **Y** 1.2	40 **Zr** 1.4	41 **Nb** 1.6	42 **Mo** 1.8	43 **Tc** 1.9	44 **Ru** 2.2	45 **Rh** 2.2	46 **Pd** 2.2	47 **Ag** 1.9	48 **Cd** 1.7						
4p													31 **Ga** 1.6	32 **Ge** 1.8	33 **As** 2.0	34 **Se** 2.4	35 **Br** 2.8	36 **Kr**
5s	37 **Rb** 0.8	38 **Sr** 1.0																
5d			57–71 1.1–1.2	72 **Hf** 1.3	73 **Ta** 1.5	74 **W** 1.7	75 **Re** 1.9	76 **Os** 2.2	77 **Ir** 2.2	78 **Pt** 2.2	79 **Au** 2.4	80 **Hg** 1.9						
5p													49 **In** 1.7	50 **Sn** 1.8	51 **Sb** 1.9	52 **Te** 2.1	53 **I** 2.5	54 **Xe**
6s	55 **Cs** 0.7	56 **Ba** 0.9																
6d			89 1.1															
6p													81 **Tl** 1.8	82 **Pb** 1.8	83 **Bi** 1.9	84 **Po** 2.0	85 **At** 2.2	86 **Rn**
7s	87 **Fr** 0.7	88 **Ra** 0.9																

POLYATOMIC MOLECULES

```
    +---+---+---+---+---+---+---+---+---+---+---+
    0.0 0.3 0.6 0.9 1.2 1.5 1.8 2.1 2.4 2.7 3.0 3.3
              Electronegativity difference
```

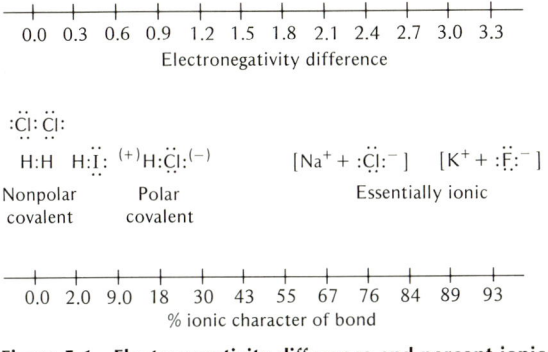

:Cl̈:Cl̈:
H:H   H:Ï: $^{(+)}$H:C̈l:$^{(-)}$ [Na$^+$ + :C̈l:$^-$]   [K$^+$ + :F̈:$^-$]
Nonpolar   Polar   Essentially ionic
covalent   covalent

```
    +---+---+---+---+---+---+---+---+---+---+
    0.0 2.0 9.0 18 30 43 55 67 76 84 89 93
              % ionic character of bond
```

Figure 5-1 Electronegativity difference and percent ionic character of chemical bonds.

types which may form between atoms whose electronegativity differences are intermediate to these two extremes? The fact is that there is no sharp distinction between bond types but a continuum, which stretches from one extreme to the other. It is helpful sometimes to refer to the percent of ionic character of a bond. When the electronegativity difference is less than 0.6, we say that the bond has less than 10 percent ionic character. For electronegativity differences of 2.0 or greater, we say that the bond has 63 percent or more ionic character. Figure 5-1 illustrates this point.

Polyatomic Molecules

Important examples of molecules in which more than one covalent bond have been formed are provided by the three common substances, water, H_2O; ammonia, NH_3; and methane, CH_4. Water is a liquid at room temperatures; ammonia and methane are gases. A single molecule of each of these consists of a central atom covalently bonded to a sufficient number of hydrogen atoms to satisfy the octet rule for the central atom. In water, the oxygen atom, with six valence electrons, bonds with two hydrogen atoms; in ammonia, the nitrogen atom, with five valence electrons, bonds with three hydrogen atoms; in methane, the carbon atom, with four valence electrons, bonds with four hydrogen atoms. Electron formulas for these three substances are

:Ö:
H H   H:N̈:H   H:C̈:H
                 H   H
Water   Ammonia   Methane

It is possible, by methods which we shall not go into at the moment, to measure certain properties of individual molecules. The most significant of these for our purposes are *polarity* and *bond angles*.

In a two-atom molecule, the presence of a *polar bond* results quite simply in a *polar molecule*. The situation is not as simple in molecules consisting of more than two atoms. In these, the geometric arrangement of the atoms may be such that the polarities of the bonds are canceled. Consider a theoretical molecule whose formula is of the A_2B form, with each A—B bond being polar. If the geometric arrangement of the three atoms is a *linear* one,

$$\overset{+}{:\!\ddot{A}\!:}\overset{-}{\ddot{B}}\overset{+}{:\!\ddot{A}\!:}$$

then the symmetry of the molecule causes the polarities of the two bonds to cancel each other and the *molecule* is completely nonpolar. If, however, the three atoms have a *triangular* arrangement

$$\overset{-}{\underset{+\quad+}{\ddot{B}}}\\:\!\ddot{A}\!:\quad:\!\ddot{A}\!:$$

there will be a negative and a positive "side" to the molecule, due to the unsymmetrical arrangement and the molecule will be polar. Information about the polarity of the molecule and of the sizes of bond angles within the molecule together with knowledge of the relative polarity of each bond is usually sufficient to develop an accurate description of the *geometry of the molecule.*

The water molecule has been found to be polar, which tells us that there is an asymmetrical arrangement of the hydrogen atoms about the oxygen atom. The H—O—H bond angle has been determined to be about 105°. On the basis of this we can visualize the H_2O molecule as a "bent" molecule.

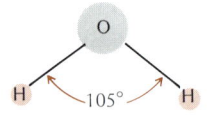

The ammonia molecule is also polar, and thus asymmetrical, with three equivalent bond angles of about 108°. This requires a three-dimensional figure for its representation, a triangular pyramid. It requires but a brief reference to geometric principles to validate this statement. A planar NH_3 molecule would restrict the three equivalent N—H—N bond angles to 120° and furthermore would yield a symmetric and nonpolar molecule.

75 THE THEORY OF HYBRID ORBITALS

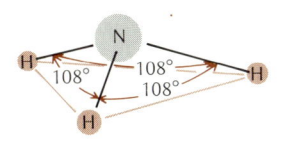

The methane molecule is nonpolar. This indicates that it must have a perfectly symmetrical geometric shape, since a consideration of the respective electronegativities of carbon and hydrogen tells us that each bond must be polar. The geometric arrangement, then, must produce a cancellation of the bond polarities. The size of the bond angles gives us the necessary clue. There are four equivalent bond angles, each 109.5°. This is the angle of the regular tetrahedron, the shape we therefore accept for the methane molecule.

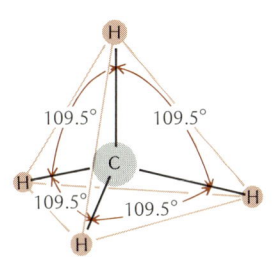

The Theory of Hybrid Orbitals

The formation of four equivalent bonds by the carbon atom, as in the methane molecule, may arouse some questions in the student's mind, especially if he has a clear picture of the electron configuration of the carbon atom:

Carbon has four valence electrons, but how can four equivalent, symmetrically directed bonds be formed by electrons in two entirely different kinds of orbitals—a spherical s orbital, and two-lobed p orbitals? We have already indicated that it is an oversimplification to consider that the atomic orbitals maintain their identity in a molecule. As this example shows, more than an oversimplification, the idea may be completely misleading.

It requires no great perception to understand that the orbitals of valence electrons in the methane *molecule*, where every electron is attracted to *each* of the five nuclei, are considerably modified as compared with those on the isolated atoms. The theory which explains the use of electrons in different types of atomic orbitals to form *equivalent* bonds is the *theory of hybridization of orbitals.* In the methane molecule, the four valence electrons of the carbon atom are considered to occupy four equivalent orbitals, formed from one *s* orbital and three *p* orbitals. These four orbitals are called *hybrid orbitals* and are given the designation of sp^3 orbitals. Each one of the four sp^3 orbitals takes one-fourth of its character (shape and distance from the nucleus) from an *s* orbital, and three-fourths of its character from the *p* orbitals. The four hybrid orbitals are completely equivalent, and bonding involving all four of them results in four equivalent, symmetrically directed bonds and a symmetric molecule. The bonds in methane are described as sp^3-*s* bonds, since each is a result of overlap of one sp^3 hybrid orbital on the carbon atom with one *s* orbital on a hydrogen atom.

The theory of hybridization can be applied to all molecules, and with some exceptions it permits us to predict the geometry of a molecule from a consideration of the original atomic orbitals. Thus, in the ammonia molecule, the nitrogen atom with the configuration

$$N \quad \underset{1s}{[\uparrow\downarrow]} \quad \underset{2s}{[\uparrow\downarrow]} \quad \underset{2p}{[\uparrow|\uparrow|\uparrow]}$$

may be considered to have four equivalent sp^3 hybrid orbitals, only three of which are used to form a bond, the fourth being occupied by the unshared pair of electrons.

$$\text{H:}\overset{..}{\underset{\underset{H}{..}}{N}}\text{:H} \quad \text{Unshared pair}$$

However, the shape of the NH_3 molecule is *not* simply three-fourths of a tetrahedron. If it were, the N—H—N angles would be 109.5°. The presence of the unshared pair of electrons in the fourth hybrid orbital has a distorting effect upon the remainder of the molecule. The unshared pair of electrons tends to occupy a larger portion of space than a hydrogen atom would, so that the remaining three angles are slightly reduced (from 109.5 to 108°).

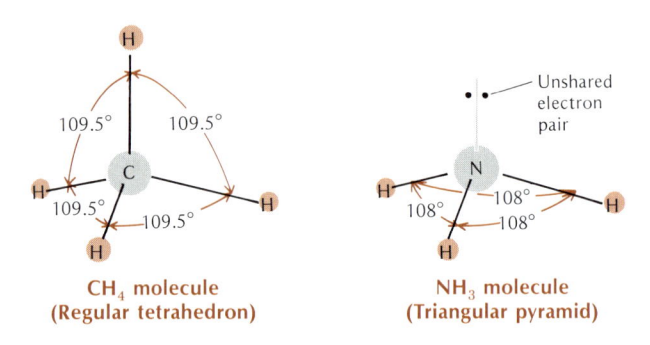

CH_4 molecule
(Regular tetrahedron)

NH_3 molecule
(Triangular pyramid)

The geometry of the water molecule could also have been predicted by the theory of hybridization. In this case, of the four sp^3 hybrid orbitals, only two are used to form bonds, and the resulting distortion of the bond angles due to the space occupied by the *two* unshared pairs is greater (109.5 to 105°). That the water molecule is planar is understandable when we refer to the geometric principle which states that three points determine a plane.

The three molecules discussed in the preceding paragraphs are representative of many similar molecules formed through utilization of sp^3 hybrid orbitals on the central atom. The geometric results are always the same: (1) formation of *four* bonds through sp^3 hybrid orbitals on the central atom produces a *tetrahedral molecule;* (2) formation of *three* bonds through sp^3 hybrid orbitals, leaving one sp^3 hybrid orbital to be occupied by an unshared pair of electrons results in a *triangular-pyramidal molecule;* (3) formation of *two* bonds, with two sp^3 hybrid orbitals being occupied by unshared pairs, results in a *bent molecule;* (4) formation of *one* bond, from *any* orbital type results in a *linear molecule*, this being the only possible space relation between two atoms. Bond angles vary, being influenced by several factors, among which are relative atomic sizes and electronegativities.

Ternary Compounds

Thus far we have discussed the structural units of only binary (two-element) compounds. There are many ternary (three-element) compounds whose structures can be understood at this time.

Two common compounds exist which consist of sodium, sulfur, and oxygen. It can be shown that each of these ionic compounds has as structural units two simple sodium ions for every dinegative ion, the dinegative ion consisting of the sulfur and oxygen atoms covalently bonded to each other. The empirical formula and electron formula for the first of these compounds, sodium sulfate, are

$$Na_2SO_4 \qquad \left[2Na^+ + \begin{array}{c} :\ddot{O}: \\ :\ddot{O}:\ddot{S}:\ddot{O}: \\ :\ddot{O}: \end{array} \right]^{--}$$

The sulfate ion illustrates a number of interesting facts:

1 The arrangement, which shows each oxygen atom bonded to the sulfur atom but no bonds between oxygen atoms, is typical. *Atoms of the same element generally do not share electrons in compounds.* Carbon is a notable exception to this generalization. The fact that a carbon atom can bond to another carbon atom in compounds gives rise to

78 DISCRETE MOLECULES AND POLYATOMIC IONS

the almost limitless possibilities of carbon-chain and carbon-ring compounds of *organic chemistry*. Carbon compounds and other exceptions will be discussed in the next chapter.

2 The total number of electrons which is necessary if each atom in the ion is to have eight electrons in its outer energy level is obtained only when the electrons lost by the metal atoms are included. (Remember the ratio of two sodium ions to one sulfate ion.)

3 The oxygen atom, with a configuration

$$O \quad \underset{1s}{[\uparrow\downarrow]} \quad \underset{2s}{[\uparrow\downarrow]} \quad \underset{2p}{[\uparrow\downarrow|\uparrow|\uparrow]}$$

has only six valence electrons. This means that the oxygen atom has contributed no electrons to the bond it has formed with the central sulfur atom. A covalent bond of this type, in which both electrons come from the same source, is called a *coordinate covalent bond*.

4 The geometry of the sulfate ion can be predicted from an application of the theory of hybridization of orbitals. The sulfur atom forms four sp^3 hybrid orbitals. Since four bonds are formed, we should expect a tetrahedral geometry and bond angles of 109.5°, and this is indeed the case.

The other sodium, sulfur, and oxygen compound is sodium sulfite, with the empirical and electron formulas

$$Na_2SO_3 \qquad \left[2Na^+ + :\overset{..}{\underset{..}{O}}:\overset{..}{\underset{..}{S}}:\overset{..}{\underset{..}{O}}:^{--} \right]$$

The sulfite ion is similar to the sulfate ion, but the fact that only three bonds are formed alters the geometry. The SO_3^{--} ion has a triangular pyramidal geometry.

Electron formulas for some other common ternary compounds are given below:

$$\left[K^+ + :\overset{..}{\underset{..}{O}}:\overset{..}{\underset{..}{Cl}}:\overset{..}{\underset{..}{O}}:^{-} \right] \qquad \left[K^+ + :\overset{..}{\underset{..}{O}}:\overset{..}{\underset{..}{Cl}}:^{-} \right]$$

$KClO_4$
Potassium perchlorate
(ClO_4^- ion has a
tetrahedral geometry.)

$KClO_3$
Potassium chlorate
(ClO_3^- ion has a
triangular-pyramidal geometry.)

TERNARY COMPOUNDS

$$\left[K^+ + :\ddot{\underset{..}{O}}:\ddot{\underset{..}{Cl}}:^- \atop :\ddot{\underset{..}{O}}: \right] \qquad \left[K^+ + :\ddot{\underset{..}{Cl}}:\ddot{\underset{..}{O}}:^- \right]$$

KClO$_2$
Potassium chlorite
(ClO$_2^-$ ion has a
bent geometry.)

KClO
Potassium hypochlorite
(ClO$^-$ ion has a linear
geometry.)

All the *binary ions* discussed thus far have been negatively charged. The number of positively charged two-element ions is much smaller. One, however, is important enough to deserve individual attention. It is the ammonium ion:

$$NH_4^+ \qquad H:\overset{H^+}{\underset{H}{\overset{..}{N}}}:H$$

The ammonium ion is formed when the ammonia molecule, which has one unshared pair of electrons, uses that pair to attach a proton. The student may here wish to review the discussion of the proton in Chap. 2. This is another example of coordinate covalence, since both the electrons for the bond are provided by the ammonia molecule. The difference in the way the bonds were formed is unimportant once the NH$_4^+$ unit is formed. All four bonds in the ion are equivalent, with the resulting expected tetrahedral structure. An example of a compound containing the ammonium ion is ionic ammonium chloride:

$$NH_4Cl \qquad \left[H:\overset{H^+}{\underset{H}{\overset{..}{N}}}:H + :\ddot{\underset{..}{Cl}}:^- \right]$$

EXCEPTIONS TO THE RULE OF EIGHT

The fact that most atoms tend to gain the stable electron configuration of a rare gas has formed the basis for much of our explanation of chemical bonding. Because for many atoms this requires eight electrons, we have called the tendency the rule of eight or the octet rule. Some molecules and ions, however, form without fulfilling this requirement.

Beryllium and boron exist in molecules in which the central atom has less than eight electrons. These are rare exceptions and can be remembered individually.

Molecules of some beryllium compounds are the only ones in which the central atom has only four electrons. For example, gaseous beryllium chloride, BeCl$_2$, is a linear, nonpolar molecule. Thus its two bonds must be equivalent and directed at 180° to each other.

BeCl₂   :Cl̈:Be:Cl̈:

Beryllium chloride

The electron configurations for beryllium and chlorine are

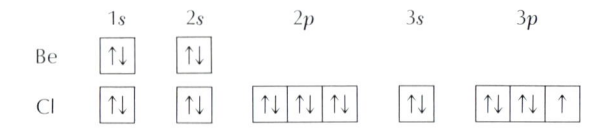

	1s	2s	2p	3s	3p
Be	↑↓	↑↓			
Cl	↑↓	↑↓	↑↓ ↑↓ ↑↓	↑↓	↑↓ ↑↓ ↑

The chlorine atom completes its octet when it gains or shares one more electron. Since only two bonds are formed, only the two 2s electrons on the beryllium atom are involved. But two bonds require two orbitals, and remembering that the two bonds must be equivalent, we find that only an explanation in terms of hybrid orbitals is satisfactory. One s and one p orbital on the beryllium atom are used to form two equivalent sp orbitals. The sp orbitals are directed at 180° to each other, thus giving a linear molecule.

Molecules in which the central atom has only six electrons are found in boron compounds:

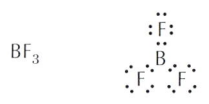

BF₃

Boron trifluoride

Boron trifluoride is a planar molecule, with the three bonds making 120° angles with each other. It is also nonpolar, which means that the three bonds are equivalent. The electron configurations for boron and fluorine are

	1s	2s	2p
B	↑↓	↑↓	↑
F	↑↓	↑↓	↑↓ ↑↓ ↑

Each fluorine atom completes its octet when it gains or shares one additional electron. The formation of three bonds will use the three valence electrons on the boron atom. Once again, the requirement that the bonds be equivalent leads to an explanation based upon hybrid orbitals. For the three bonds, one s and two p orbitals on the boron atom are used to form three equivalent sp^2 orbitals. The sp^2 orbitals are directed in the same plane, making angles of 120° with each other.

There are also molecules and ions in which the central atom has acquired *more* than eight electrons. This phenomenon is referred to as *expansion of the octet*. Since according to the theory of hybridization of orbitals each bond formed must utilize an available orbital on the central atom, only atoms with available *d* orbitals can undergo this expansion. Thus elements which can act as the central atom in a structure of this kind are only those found in period 3 or lower in the periodic table. A structure in which the central atom has ten electrons, and thus utilizes five orbitals, is a sp^3d hybrid, since one *s*, three *p*, and one *d* orbitals on the central atom have been hybridized to form five equivalent sp^3d orbitals. Twelve electrons indicate an sp^3d^2 hybrid, and fourteen, a sp^3d^3 hybrid. The geometries of these five-, six-, and seven-bonded units become quite complex. A few illustrations should suffice.

PF_5, phosphorus pentafluoride:
Ten electrons on central atom
An sp^3d hybrid
Five bonds formed

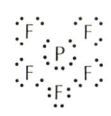

Five equivalent bonds give this molecule the geometric shape of a triangular bipyramid:

SF_6, sulfur hexafluoride:
Twelve electrons on the central atom
An sp^3d^2 hybrid
Six bonds formed

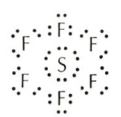

Six equivalent bonds give this molecule the geometric shape of a regular octahedron. The ions SiF_6^{--} and PCl_6^- also have octahedral shapes.

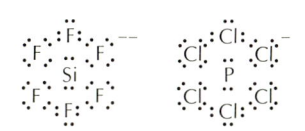

As with molecules and ions which do follow the octet rule, many molecules and ions with expanded octets are formed as a result of hybridization of orbitals but with one or more of the hybrid orbitals occupied by unshared electron pairs.

BrF_5, bromine pentafluoride:
Twelve electrons on central atom
An sp^3d^2 hybrid
Only five bonds are formed

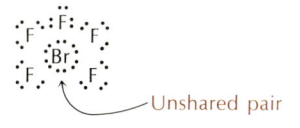

Unshared pair

DISCRETE MOLECULES AND POLYATOMIC IONS

XeF_4, xenon tetrafluoride:
 Twelve electrons on central atom
 An sp^3d^2 hybrid
 Only four bonds are formed

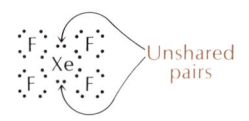

Multiple Covalent Bonds

It is frequently the case that two atoms share more than one pair of electrons—but in the same bonding direction. An illustration of this type of bond can be found in carbon dioxide, CO_2. Since both carbon and oxygen are nonmetals, the substance is molecular. The molecule of CO_2 has been shown to be linear and to have bond angles of 180°. There are only 16 valence electrons for the structure, 6 each from the two oxygen atoms, and 4 from the carbon atom. The electron formula which meets all these requirements and satisfies the octet rule for each atom is

$$\ddot{\underset{..}{O}}::C::\ddot{\underset{..}{O}}$$

The type of bond illustrated here in which two atoms share *two pairs* of electrons is called a *double bond*. There are two double bonds in the CO_2 molecule. According to the theory of hybrid orbitals, the fact that the molecule is linear implies that the two C—O bonds are formed from two *sp* hybrid orbitals on the carbon atom. The first, or *sigma, bond* of each double bond in carbon dioxide is, in fact, formed in this way. Together, these two sigma bonds are responsible for the geometry of the molecule. We shall take up the nature of the second, or *pi, bond* of the double bond in the next chapter.

The electron formula for carbon monoxide, CO, requires a triple bond:

$$:C:::O:$$

The CO molecule is also linear, with the first, or sigma, bond of the triple bond being formed from an *sp* orbital on the carbon atom. The other *sp* orbital is occupied by an unshared pair of electrons. The second and third bonds of the triple bond are both pi bonds.

We have not included *bond lengths* or *bond strengths* in our discussion of properties of molecules up to this time. These are important measurable properties. For the present we shall say only that these properties provide further evidence for the nature of the bonds in carbon dioxide and carbon monoxide. The double bond is both stronger and shorter than the single bond, and the triple bond is even stronger and shorter.

Resonance

An electron formula for the molecule of sulfur dioxide, SO_2, devised *without* information about the properties of the molecule might be drawn

$$\ddot{\underset{..}{O}}::\ddot{S}:\ddot{\underset{..}{O}}:$$

RESONANCE

This formula uses the correct number of valence electrons and furnishes each atom with a complete octet. But the representation is not acceptable when we learn that the molecule is nonlinear and polar and that the two S—O bonds are equivalent! Since we cannot show two "one-and-a-half" bonds, which is what seems to be required, the best we can do to represent the SO$_2$ molecule is to draw two equivalent forms, implying that the actual condition is an average or *hybrid* of the two:

$$:\ddot{O}\!:\!\!\overset{\ddot{S}}{}\!\!:\ddot{O}\!: \longleftrightarrow :\ddot{O}\!:\!\!\overset{\ddot{S}}{}\!\!:\ddot{O}\!:$$

The sulfur dioxide molecule is called a *resonance hybrid*.

Examples of resonance hybrids can be found among ions as well as molecules. The nitrite ion, NO$_2^-$, which is present in such ionic compounds as sodium nitrite, NaNO$_2$, is similar in its electron formula to that of the SO$_2$ molecule, since the total number of valence electrons available for the structure is the same.

$$\text{NaNO}_2 \quad \left[\text{Na}^+ + \overset{\cdot\cdot}{\text{N}}\!::\!\ddot{\text{O}}\!:^- \longleftrightarrow \overset{\cdot\cdot}{\text{N}}\!:\!\ddot{\text{O}}\!:^- \right]$$

The sulfur trioxide molecule, SO$_3$, is triangular in shape and contains three equivalent bonds whose strength is intermediate between a single and a double bond. The electron formula for this resonance hybrid is

$$\begin{array}{ccc} :\ddot{O} \quad :\ddot{O}: & :\ddot{O}: \quad :\ddot{O}: & :\ddot{O}: \quad \ddot{O}: \\ S & \longleftrightarrow \quad S \quad \longleftrightarrow & S \\ :\ddot{O}: & \ddot{O}. & :\ddot{O}: \end{array}$$

The carbonate ion, CO$_3^{--}$, which is present in such ionic compounds as sodium carbonate, Na$_2$CO$_3$, is similar in electron formula to the SO$_3$ molecule, since it too has 24 valence electrons in its structure:

$$\text{Na}_2\text{CO}_3 \quad \left[2\text{Na}^+ + \begin{array}{c}:\ddot{O}\quad:\ddot{O}:\\ C \\ :\ddot{O}:\end{array}^{--} \longleftrightarrow \begin{array}{c}:\ddot{O}:\quad:\ddot{O}:\\ C \\ \ddot{O}.\end{array}^{--} \longleftrightarrow \begin{array}{c}:\ddot{O}:\quad\ddot{O}:\\ C \\ :\ddot{O}:\end{array}^{--} \right]$$

The similarity existing between the electron formulas of the nitrite ion and the SO$_2$ molecule and between the carbonate ion and the SO$_3$

molecule is given a name. *Molecules or ions with identical electron structures are said to be isoelectronic.* Another example of this is the relationship between the nitrogen molecule and the CO molecule:

$$:N:::N: \quad :C:::O:$$

Although electronic structures for these two molecules are identical, it should be obvious that there will be a marked difference in at least one other molecular property. The nitrogen molecule is nonpolar, while the carbon monoxide molecule exhibits polarity due to the electronegativity difference between carbon and oxygen.

THE OXYGEN MOLECULE

Bonding in the oxygen molecule is less simple than in other diatomic molecules considered up to now. Properties of the molecule imply both multiple bonding and deviations from the rule of eight. The 12 electrons might seem to indicate a double bond:

$$:\overset{.}{O}::\overset{.}{O}:$$

The bond between the two atoms is stronger than a single bond, and so this formula has some relation to the structure. However, oxygen has the property of being *paramagnetic,* or of being slightly attracted to a magnet. This property is a direct result of the presence of *unpaired electrons* in a molecule. Paired electrons, the student will remember, have opposite spins so that they cancel each other out so far as effect upon the molecule goes. But unpaired electrons have parallel spin and thus impart a very slight magnetic field. The property of paramagnetism, then, requires the consideration of the following representation for the O_2 molecule:

$$:\overset{..}{\underset{.}{O}}:\overset{..}{\underset{.}{O}}:$$

This formula accounts for the paramagnetism of the molecule but not for the strength of the bond. Thus neither of the above formulas is completely satisfactory. The possibility of resonance is strongly suggested here and the O_2 molecule is often represented as a resonance hybrid.

$$:\overset{.}{O}::\overset{.}{O}: \longleftrightarrow :\overset{..}{\underset{.}{O}}:\overset{..}{\underset{.}{O}}:$$

QUESTIONS

Electron Formula Summary

It will be helpful to summarize the rules and generalizations for writing the electron formula for a given substance.

1. Decide whether the compound is essentially ionic or essentially covalent. This usually requires no more than identification of the elements as metals or nonmetals. If necessary, electronegativities may be considered.
2. If the substance is ionic, use brackets and separate the positive ion from the negative ion with a plus sign. Charges on the ions must be determined. For monatomic ions, review the discussion of the formation of ions in Chap. 4. Charges on polyatomic ions should be noted and remembered.
3. In writing electron formulas for a molecule or polyatomic ion:
   a. Determine the total number of electrons available for the structure by adding valence electrons of all atoms involved plus (in the case of a covalently bonded ion) any electrons obtained from the metal atoms of the compound.
   b. Arrange the atoms so that atoms of the same element are not bonded to each other.
   c. Place the electrons around the atoms in such a way that each is supplied with eight. If the number of electrons is *insufficient* to provide all atoms with an octet while using single bonds, incorporate double or triple bonds as needed (exceptions: beryllium and boron compounds).
   d. When the simplest electron formula shows a multiple bond in one of two or more equivalent positions, use the resonance-hybrid representation.
   e. When the total of electrons available is *more* than required to provide each atom with eight electrons, it indicates that the central atom has undergone expansion of the octet.

Questions

1. a. Use Table 5-1 to arrange the following bonds in order of increasing polarity:

C—Br	P—O	Na—Cl
S—O	Ca—O	C—H
Si—Cl	Ba—Cl	C—F

   b. Which of these bonds will be essentially ionic (63 percent or more ionic character)? Which will be essentially covalent (less than 10 percent ionic character)?

2. Locate the relative positions of the elements magnesium and barium in the periodic table. Which will have the larger atom? The higher ionization energy? The larger electron affinity? The larger electro-

negativity? In the same way, compare the pair of nonmetals chlorine and iodine and then the elements calcium and bromine.

3 a Oxygen reacts to form compounds with each of the first four elements of group VA. In which of these compounds would the bonds be most ionic? In which the most covalent?

b Calcium forms a compound with each of the elements in group VIIA. Which of these compounds is most ionic? Which is most covalent?

c Which binary compounds formed between nitrogen and the elements of group IA would be the most covalent? The most ionic?

4 a The bond angles in the molecule of H_2S have been found to be 92°20'. Would you expect this molecule to be polar or nonpolar? Explain your answer.

b Bond angles in PH_3 have been found to be 93°50'. Would you expect this molecule to be polar or nonpolar? Why?

c The BF_3 molecule has bond angles of 120°. What geometry would you predict for this molecule? Would it be polar or nonpolar?

d The $BeCl_2$ molecule is nonpolar. What does this tell about its bond angles?

e The CCl_4 molecule is nonpolar. Would you expect CCl_3F to be polar or nonpolar?

5 Consider the five hypothetical elements, V, W, X, Y, and Z, with one, two, four, six, and seven valence electrons, respectively, located in the second or lower periods in the periodic table.

a Write electron formulas for monatomic ions of V, W, Y, and Z.

b Classify the following *bonds* as probably essentially ionic or probably essentially covalent:

W—Y X—Z X—Y
W—Z V—Y V—Z
Z—Z

c Write complete electron formulas for two ionic and two covalent compounds which might be formed from these elements.

6 Consider the hypothetical elements R and T. Element R is in group VIA of the periodic table, and T is in group VIIA. What kind of bond (ionic or covalent) is formed when these elements react? Write the electron formula for the compound formed.

7 Consider the hypothetical elements with symbols of A and B. Element A has an atomic number of 7, and B has an atomic number of 17. Would a compound formed between these two elements be essentially ionic or essentially covalent? Write the electron formula for the compound which would be formed.

8 Give the formulas for each of the following:

a Three monatomic ions   **b** One diatomic ion
c Two triatomic ions   **d** Three tetratomic ions

87 QUESTIONS

  e Two pentatomic ions
  f Four diatomic molecules
  g Three triatomic molecules
  h Three tetratomic molecules
  i One pentatomic molecule
  j Two polyatomic ions in which a sulfur atom is the central atom
  k Four polyatomic ions in which a chlorine atom is the central atom
  l Two polyatomic ions in which a nitrogen atom is the central atom
  m One polyatomic ion in which a phosphorus atom is the central atom
  n Two molecules in which the central atom is a carbon atom
  o Two molecules in which the central atom is a sulfur atom
  p One molecule in which the central atom is a nitrogen atom
  q Three monovalent positive ions
  r Two monovalent negative ions
  s Two divalent positive ions
  t Five divalent negative ions

9 a Electron formulas for the seven units in the table below have been given in the text. Fill in the table as has been done for the first two. (In counting "number of bonds," consider double or triple bonds each as simply *one* bond.)

Unit	No. of Atoms per Unit	Ionic Charge (If Any)	Total No. of Valence Electrons	No. of Bonds	Is Resonance Required?	Is Octet Expansion Required?	Electron Formula
NH_3	4	0	8	3	No	No	H:N:H with H above
CO_3^{--}	4	−2	24	3	Yes	No	(three resonance structures shown)
SO_4^{--}							
Ca^{++}							
NO_3^-							
H_2O							
SiF_6^{--}							

88 DISCRETE MOLECULES AND POLYATOMIC IONS

b Construct a similar table for the following 11 units. By using the method summarized on page 85 you should be able to provide all information in the table, including electron formulas:

H_2S CF_4 PCl_6^-
HBr SiF_4 PH_3
PO_4^{3-} BrO_3^- CN^-
IO_3^- OCN^- (C as central atom)

10 Draw electron formulas for the following:
   a NH_4Br **b** $CaSO_3$ **c** Na_3PO_4
   d K_2CO_3 **e** NH_4NO_3 **f** $Ca(BrO_3)_2$

11 For each of the following ions, give the formula of a neutral molecule which corresponds to the ion in both electron formula and geometry:
   a SO_4^{--} **b** NO_2^- **c** CO_3^{--}
   d CN^- **e** NO_3^- **f** PO_4^{3-}
   g NH_4^+ **h** ClO_3^- **i** ClO_4^-
   j ClO_2^- **k** OCN^-

12 Among the molecules and ions discussed so far:
   a The electron formulas for five molecules and for five ions show multiple bonds. Identify them.
   b The electron formulas for two molecules and three ions require resonance. What are they?
   c The electron formulas for four molecules and two ions show octet expansion. Can you list them?

13 Give electron formulas for the ions and molecules below. Formulas for these have not been given, but the student may find it helpful to consider first what more familiar ion or molecule may be isoelectronic with the given one. For instance, H_2S is isoelectronic with H_2O, since O and S both have six valence electrons. The electron formula for H_2S, then, is H:S:H (with lone pairs on S).

CCl_4 Ba^{++} PH_4^+
H_2Se IF_5 BrO^-
I_2 HI Rb^+
C_2^{--} SeF_6 SCN^-
I^- I_2 Li^+
P^{3-} CS_2 CBr_4
F_2 Br_2 BCl_3
PCl_5 O^{--} (oxide ion) O_2^{--} (peroxide ion)

14 Nitrogen has a relatively high electronegativity, but the N_2 molecule is quite inert. Explain.

15 Account for the fact that H_2O is a polar molecule and CCl_4 is not.

16 Write the electron formulas for BF_3 and NF_3. Write the formula for the compound you predict will be formed if one molecule of BF_3 reacts with one molecule of NF_3.

QUESTIONS

17 Why is it necessary to speak of hybridization of bonds when discussing the bonds in CH_4, in NH_3, in BF_3?
18 What is the geometry of the NH_4^+ ion?
19 Assume sp^3 hybridization for all the molecules and ions listed below. Indicate the molecular shape of each:
    a HF   **b** PH_3 **c** H_3O^+
    d SO_3^{--} **e** SO_4^{--} **f** SiH_4
    g H_2S **h** CBr_4
20 Can a filled orbital take part in the bonding process? Explain.
21 What information does the formula of a substance give that allows us to predict whether it will be predominantly ionic or predominantly covalent?
22 Both BF_3 and SO_3 have 24 valence electrons involved in the molecular structure but utilize them in different ways. Explain.
23 If an even number of oxygen atoms is bonded to a central atom with an odd number of valence electrons, the unit formed is always an ion rather than a molecule. Why?

CHAPTER 6 POLYMERIC MOLECULES AND IONS

Giant molecules such as those of the genetically important deoxyribonucleic acids (DNA) present a challenging problem in structure determination. The key to the structure of the molecule represented here was the discovery of its "double helix" geometry by Watson and Crick. (Ealing Corporation, Cambridge, Mass.)

CHAPTER 6 POLYMERIC MOLECULES AND IONS

Molecules and ions described in Chap. 5 are discrete structural units. This refers to the fact that these units are separate and distinct from one another, with the molecule consisting of exactly the number of atoms shown in the formula. Other ions and molecules exist which, through repetition of all or part of the structural unit, may have large chainlike, sheetlike, or three-dimensional forms. These units are called extended or *polymeric* molecules and ions. The term polymer is used to refer to very large molecules or ions formed by the repeated combination of simpler units. Formulas for substances of this type frequently contain an indefinite subscript. The formula for the ionic compound ammonium polysulfide, $(NH_4)_2S_x$, indicates the extended nature of the dinegative ion, and the formula for molecular metaphosphoric acid, $(HPO_3)_n$, indicates the repetition of the whole formula unit in the molecule.

A bond formed between two atoms of the same element is known as a *homonuclear bond*. The single covalent bond of the diatomic molecule of chlorine or of any of the other elemental gases is of this type. Certain of the nonmetal atoms have the ability to form more than one homonuclear bond per atom, with a resulting buildup of long chainlike or more complicated molecules. Carbon exhibits this property to a greater degree than any other element, but homonuclear molecules and ions of silicon, phosphorus, sulfur, and selenium are important.

Polymeric molecules and ions are also formed by the repetition of structural units which do not involve homonuclear bonds. An alternating element-to-oxygen-to-element bonding is usually characteristic. Most of the nonmetals form some compounds of this type, silicon, boron, and phosphorus compounds being the most numerous and most important.

Homonuclear Bonding

Of all of the elements only the rare gases, hydrogen, nitrogen, oxygen, and the halogens consist of discrete molecular units in the solid state. All other nonmetals and metalloids consist of larger polymeric units in the solid state. These units are held together by homonuclear bonds. We shall consider the structure of metals in the next chapter.

SULFUR, SELENIUM, TELLURIUM

Sulfur, element 16, is in group VIA of the periodic table. The atom of sulfur has an electron configuration with an s^2p^4 ending. The common simple ion of sulfur is the dinegative sulfide ion, S^{--}, formed when the atom completes its octet by gaining a pair of electrons.

At room temperature sulfur is a yellow solid. A careful examination of the particle nature of this solid reveals that there are two distinct forms of solid sulfur. Both forms are crystalline, but one consists of *orthorhombic crystals* and the other of *monoclinic crystals* (see Fig. 6-1).

POLYMERIC MOLECULES AND IONS

Orthorhombic sulfur

Monoclinic sulfur

Orthorhombic crystal

Monoclinic crystal

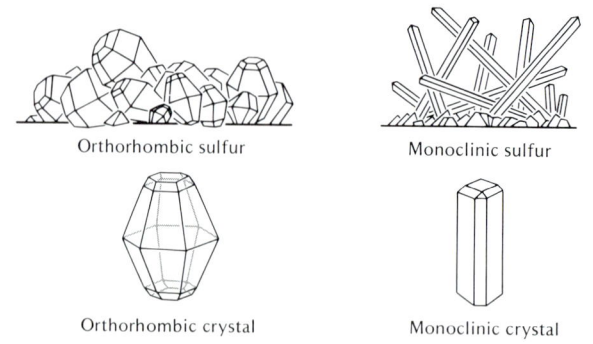

Figure 6-1 Allotropic forms of sulfur.

The observation of how heat affects liquid sulfur provides evidence that sulfur exists in two distinct liquid forms, depending upon the temperature. When solid sulfur is melted, it forms a light yellow liquid. As this liquid is heated, it darkens and becomes more and more viscous, until at 180 to 200°C these properties reach a maximum and the sulfur is a dark, sticky mass. When heating is continued above this temperature, the viscosity of the liquid steadily decreases, until at 444.6°C, it boils.

Density determinations upon sulfur vapor at various temperatures reveal that in the vapor form too sulfur exists in a number of distinct forms.

The property of an element of existing in more than one crystalline modification or in more than one molecular form is known as *allotropy*, and the various forms are called *allotropes*.

The property of the sulfur atom responsible for the existence of the several molecular forms is its ability to form homonuclear bonds. The molecular form of solid sulfur is an eight-atom ring. Each atom completes its octet by sharing a pair of electrons with the next atom (see Fig. 6-2). Each of the two crystalline forms of sulfur consists of these eight-member

Figure 6-2 The S_8 molecule.

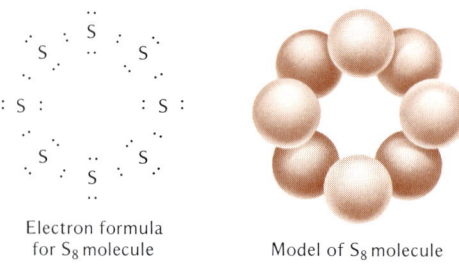

Electron formula for S_8 molecule

Model of S_8 molecule

rings, with the arrangement of the rings in the crystal being of two different patterns.

The yellow-colored liquid sulfur obtained when the solid is melted still consists of these eight-member rings. The heat energy has been utilized in overcoming the intermolecular forces which maintained the crystal structure. As the liquid is heated to higher temperatures, energy is utilized in breaking the ring open. The sulfur chains thus formed have terminal atoms which are electron-deficient; that is, they do not have completed octets. These are very reactive and tend to attach to other chains in order to share electrons. The more rings that break open, the longer the chains become, with resulting intertwining, interference, and increased viscosity. At temperatures of 180 to 200°C, some of these chains are made up of literally *thousands* of sulfur atoms. Above 200°C, all rings have been broken open, and heat energy begins to break apart the chains, until, when the liquid boils at 444.6°C, it consists almost entirely of S_8 molecules.

Continued heating of the vapor phase results in continued chain breaking, so that S_8, S_6, S_4, S_2, and S molecules are found in sulfur vapor.

Sulfur also forms extended ions. When free sulfur is dissolved in a concentrated sulfide solution, *polysulfide* ions are formed. Apparently the sulfide ion breaks open the S_8 sulfur ring, forming an S_9^{--} ion, which in turn splits and combines with more sulfide ions. Being dinegative, each of these units has a completed octet.

Selenium and tellurium, elements directly beneath sulfur in the periodic table, form many molecules and ions analogous to those of sulfur.

Although homonuclear bonds involving oxygen do exist at room temperatures, the longest chain of oxygen atoms known until recently was three (O_3). Chemists are devoting much time to chemical reactions that take place at low temperatures, and as a result compounds have been prepared at low temperatures that do not exist at ordinary temperatures. In 1966 the preparation of O_5F_2 and O_6F_2 at $-213°C$, with electron formulas of :F̈:Ö:Ö:Ö:Ö:F̈: and :F̈:Ö:Ö:Ö:Ö:Ö:F̈:, was announced. Both these compounds decompose at temperatures above $-183°C$.

PHOSPHORUS, ARSENIC, AND ANTIMONY

Phosphorus has three allotropes in the solid form, one in the liquid form, and at least two in the vapor form. The three solid forms are known as *red phosphorus, white phosphorus,* and *black phosphorus.*

The phosphorus atom has an electron configuration with an ending of s^2p^3. By forming a covalent bond with each of three other phosphorus atoms, each atom completes its octet. The molecule of white phosphorus is a simple tetrahedron of four atoms (see Fig. 6-3). Black phosphorus has a layered structure. Each atom is attached to three other atoms as

96 POLYMERIC MOLECULES AND IONS

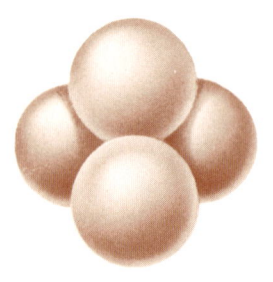

Figure 6-3 The P_4 molecule.

shown in Fig. 6-4. Red phosphorus is the most familiar form of the element. Its structure has not yet been clarified but is polymeric in nature.

Liquid phosphorus consists of P_4 molecules, as does the vapor at low temperatures. At higher temperatures the predominating form is a P_2 molecule.

Arsenic and antimony, elements directly beneath phosphorus in the periodic table, exhibit molecular forms similar to that of phosphorus.

CARBON, SILICON

The property of forming a homonuclear bond is exhibited to a greater degree by the carbon atom than by atoms of any other element. Elementary carbon exists in two allotropic forms. One of these is the diamond crystal, a transparent, stable crystal, noted for its properties of hardness and brilliance when properly cut and polished. The other allotrope of carbon is graphite. Familiar forms of graphitic carbon are charcoal, lampblack, coke, and the graphite of the so-called "lead" pencil.

The diamond crystal consists of carbon atoms, each covalently bonded to four other carbon atoms, which form the vertices of a regular tetrahedron. Because the bonding is repeated without interruption throughout the solid, the entire crystal is considered to be the molecule (see Fig. 6-5). The force between the carbon atoms is the strongest of all chemical bonds, the covalent bond. Because of this, diamond is not only one of the hardest substances known but virtually impossible to melt. Melting, as we shall see in Chap. 8, would necessitate breaking the covalent bonds.

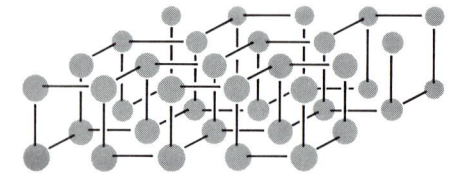

Figure 6-4 The structure of black phosphorus.

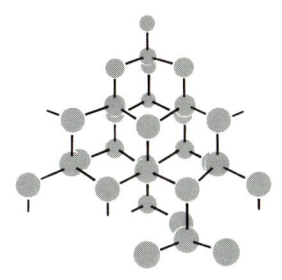

Figure 6-5 The structure of diamond.

Not only are energy requirements for this exceedingly high, but once the bonds are broken, the substance is no longer diamond, for it is the structure of diamond which is responsible for its properties!

The properties of the graphitic form of carbon are so dissimilar to those of the diamond that it seems impossible that the two are merely different forms of the same substance. Graphite consists of two-dimensional layers in which each carbon atom is bonded to three other carbon atoms by three bonds, each of which is stronger than a single bond but weaker than a double bond, a situation which must be depicted by a resonance formula (Figs. 6-6 and 6-7). There is no covalent bond *between* the layers of graphite. The force holding one layer to another is an attraction known as *van der Waals forces* (an explanation of their nature will be found in Chap. 7). For the time being we note that they are among the weak intermolecular forces, while the covalent bond is one of the strongest bonding forces. The term graphite is ordinarily applied only to massive crystals of graphitic carbon. Lampblack, boneblack, and charcoal are more frequently termed "amorphous carbon" (amorphous meaning "without form") due to their apparent lack of crystalline structure.

The study of *organic chemistry* is the study of the hundreds of thousands

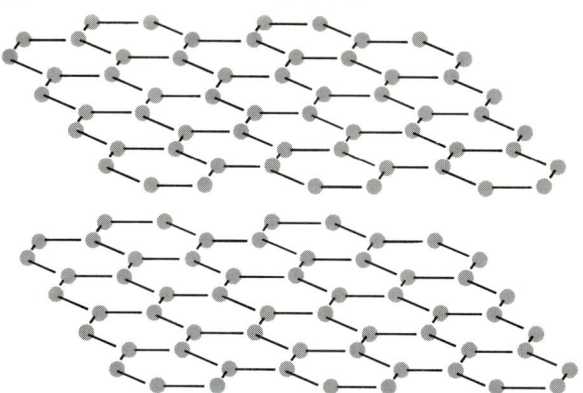

Figure 6-6 The structure of graphite.

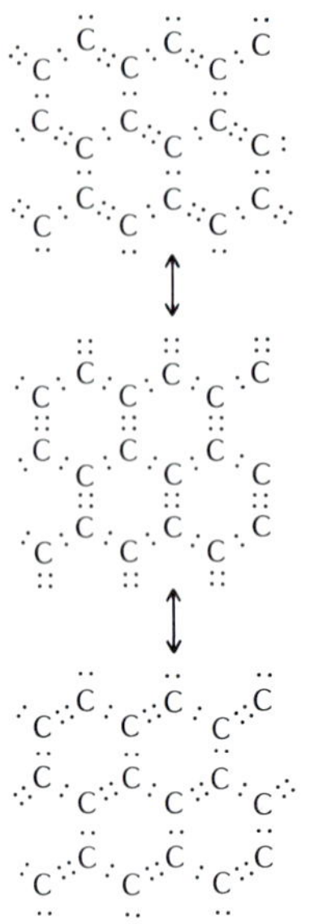

98 POLYMERIC MOLECULES AND IONS

Figure 6-7 Resonance formula for a portion of the graphite molecule.

of compounds containing homonuclear carbon-to-carbon bonds. Let us examine some of the factors which are responsible for the multiplicity, variety, and complexity of carbon compounds.

We shall make use of so-called *structural formulas* for these compounds. Structural formulas are similar to electron formulas except that they employ a line to represent the covalent bond and do not show unshared valence electrons. Formulas for the compound methane, CH_4, are

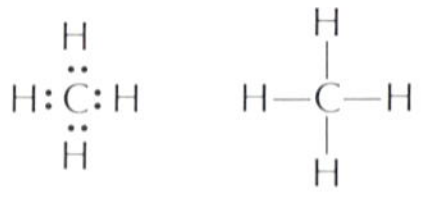

Electron formula   Structural formula

The number of carbon-to-carbon bonds which exists in a molecule may be as small as two or three, as in the molecules of ethane, C_2H_6, and propane, C_3H_8, or as large as a thousand.

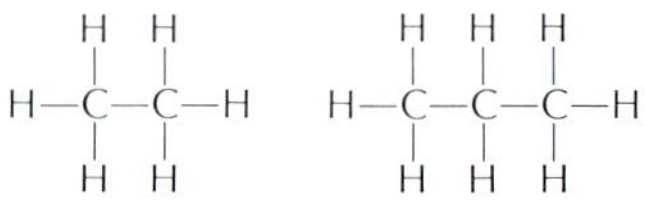

Ethane   Propane

The basic structure of the molecule may be a ring, as well as a chain, for example cyclopentane, C_5H_{10}, and cyclohexane, C_6H_{12}:

HOMONUCLEAR BONDING

Cyclopentane   **Cyclohexane**

One disadvantage of the structural formula, and likewise of the electron formula, is that it oversimplifies the geometry of the molecule. The four single bonds of a carbon atom are tetrahedrally oriented, making angles of 109.5° with each other, rather than 90° as the formula seems to indicate. For this reason, reference to three-dimensional models or at least to diagrams containing some perspective is wise. Ball-and-stick models, in which a grey ball represents a carbon atom, a colored ball a hydrogen atom, and a stick a covalent bond, represent the molecules of ethane and propane in this way:

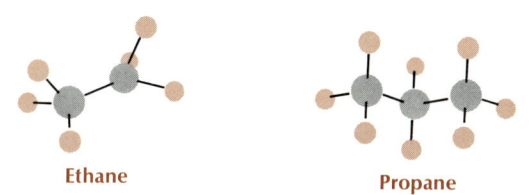

Ethane   **Propane**

These models clearly show that the term *straight-chain compounds* is a misnomer. Because of the angles at which the bonds form, the chain is forced into a zigzag pattern.

These models can also be used to demonstrate an important property of the single bond, *freedom of rotation*. Because of this property, all the representations below are of the *same* molecule. The variation has been brought about by rotation about a single bond—something which is not only permitted but is in fact taking place in the molecule all the time.

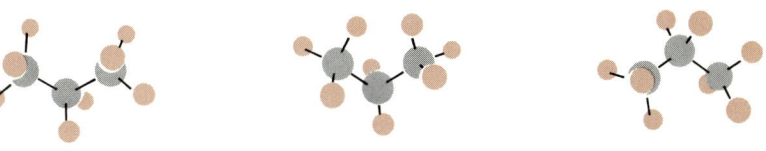

The propane molecule

More complexity is introduced to the field of carbon compounds by the fact that one or more of the carbon-carbon bonds may be a double or a triple bond. This gives two simple compounds, ethene, C_2H_4, and ethyne, C_2H_2, each with one multiple bond per molecule, and hundreds

100 POLYMERIC MOLECULES AND IONS

of larger molecules which contain both kinds of bonds. The common name for ethene is ethylene. Ethyne is more commonly called acetylene.

$$\begin{matrix} H \\ \end{matrix} C=C \begin{matrix} H \\ \end{matrix} \qquad H-C\equiv C-H$$

Ethene (ethylene)   **Ethyne (acetylene)**

A double or a triple bond in a molecule affects its geometry. Consider molecules of ethane, C_2H_6, and ethene, C_2H_4.

$$H-\underset{H}{\overset{H}{C}}-\underset{H}{\overset{H}{C}}-H \qquad \begin{matrix} H \\ \end{matrix} C=C \begin{matrix} H \\ \end{matrix}$$

Ethane   **Ethene**

Each carbon atom of the ethane molecule has four tetrahedrally oriented bonds. Three of these are sp^3-s bonds to hydrogen atoms and one is an sp^3-sp^3 bond to the other carbon atom. Thus the molecule has a geometric shape corresponding to two tetrahedra connected through a vertex on each (Fig. 6-8a).

Direct experimental evidence shows the ethene molecule to have a flat,

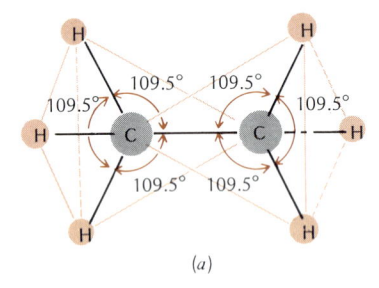

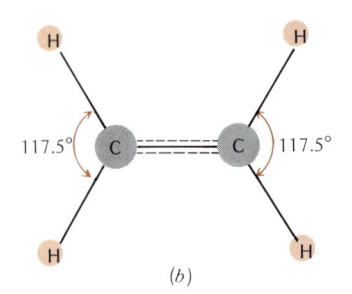

Figure 6-8 The geometries of (a) ethane and (b) ethene. A pi bond is represented by two dotted lines, one above and one below the solid line which represents a sigma bond.

101 HOMONUCLEAR BONDING

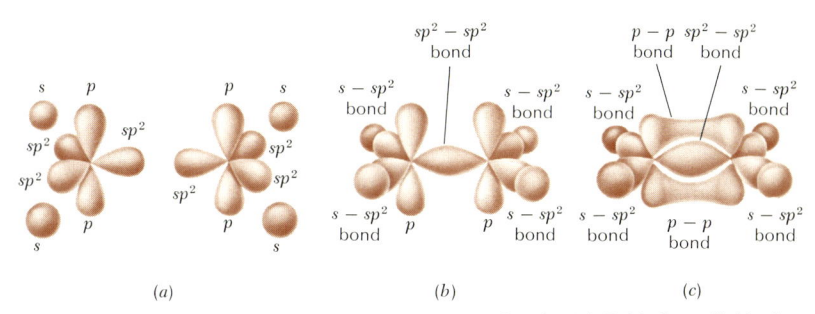

Figure 6-9 Orbital relationships in the ethene molecule. (a) Orbitals available for bonding on two carbon atoms and four hydrogen atoms; (b) formation of the five sigma bonds of the ethene molecule; (c) formation of the pi bond between the two carbon atoms in the ethene molecule.

or planar, shape. Each carbon atom has formed bonds in three directions only. Bond angles are close to 120°. In terms of hybrid orbitals, this implies sp^2 hybridization. What has become of the fourth carbon atom orbital and its electron? Figure 6-9 is an attempt to give visual meaning to the orbital relationships in ethene. In (a) the orbitals available for bonding are shown. Each of the four hydrogen atoms has one s orbital, containing one electron, and each of the two carbon atoms has three sp^2 hybrid orbitals, each orbital containing one electron, and one pure p orbital at right angles to the plane of the sp^2 hybrid orbitals, also containing an electron. (b) shows the formation of the five sigma bonds of the molecule, four of which are formed by each hydrogen s orbital overlapping a different carbon sp^2 orbital, and the fifth bond formed by an sp^2 orbital on one carbon overlapping an sp^2 orbital on the other carbon. In (c) the additional bonding between the two carbon atoms due to the overlap of the p orbitals with an edgewise orientation is shown. This additional bonding is what is termed a *pi bond*, and the sum of the sigma and pi bonding between the two carbon atoms is what is known as a double bond. An important characteristic of the double bond is due to the fact that the orbitals used in the formation of the second, or pi, bond lie at right angles to the three hybrid sp^2 bonds. This pi bond exists only as long as the orbitals are lined up with each other so that they overlap. The result is that there can be no rotation about this bond. A one-half twist of the molecule will separate the orbitals used in forming the pi bond or, in other words, will rupture the bond and destroy the identity of the molecule. The molecule of ethene, then, has no permitted rotation of one carbon atom with respect to the other.

Additional variation among carbon compounds is due to the existence of the *benzene ring*. The previously mentioned ring compounds, cyclopentane and cyclohexane, differ from normal pentane and normal hexane

102 POLYMERIC MOLECULES AND IONS

only in that the terminal carbon atoms are bonded to each other instead of to two hydrogen atoms. Consider normal hexane and cyclohexane:

Normal hexane   **Cyclohexane**

Molecular formulas and molecular weights differ slightly for these two compounds, which as might be expected, also differ somewhat in properties.

The benzene ring is also a six-carbon ring, but it differs markedly in all its properties from cyclohexane. The structural formula for benzene is

Benzene

Familiarity with resonance formulas by this time should permit the reader to make the correct interpretation of this formula. It represents a molecule with the molecular formula C_6H_6. This molecule contains six single carbon-to-hydrogen bonds and six carbon-to-carbon bonds which are something more than single bonds and something less than double bonds (Fig. 6-10).

Experimentation shows the benzene molecule to have an essentially

Figure 6-10 The geometry of the benzene molecule. A pi bond is represented by two dotted lines, one above and one below the solid line which represents a sigma bond.

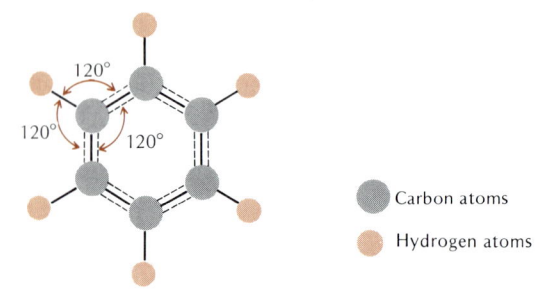

● Carbon atoms

● Hydrogen atoms

103 HOMONUCLEAR BONDING

planar geometry. From this and from other evidence, the bonding in the benzene molecule is explained as three sp^2 hybrid bonds formed by each carbon atom, one with a hydrogen atom and the other two with carbon atoms. This leaves each carbon atom with one unused p orbital and one unshared electron. As in the ethene molecule, these are used to form pi bonds, the equivalent of three pi bonds per molecule. Since all carbon-to-carbon bonds in the benzene ring are equivalent, these pi bonds must be considered to be shared evenly all around the ring.

The resonance formula is cumbersome, so that a special symbol for the benzene ring has been adopted which consists of a hexagon with a circle within, to represent the pi bonds.

The benzene ring is a particularly stable arrangement and persists through many modifications. The following few examples merely suggest the many types of *aromatic compounds,* as compounds containing the benzene ring are called.

Benzene, C_6H_6

Toluene (methylbenzene), $CH_3C_6H_5$

Phenol (hydroxybenzene), C_6H_5OH

Aniline (aminobenzene), $C_6H_5NH_2$

Hydrocarbons with more than one benzene ring per molecule also exist. Naphthalene, $C_{10}H_8$, one of these, is commonly used as moth balls.

Naphthalene

One of the major health problems of the world is cancer. Many substances whose molecular form consists of several benzene rings fused together, as in naphthalene, are known as *carcinogens,* or cancer-producing substances. Chemists have helped to establish the fact that there is a causal relationship between smoking and lung cancer. That adults continue to smoke and that young people persist in taking up smoking in spite of this established fact illustrates the contrariness of the human species.

POLYMERIC MOLECULES AND IONS

Although it is firmly established that smoking and lung cancer are directly related, the compound or compounds in cigarets or cigaret smoke that cause the cancer have not yet been identified. One possibility is the hydrocarbon benzpyrene, $C_{20}H_{12}$:

Benzpyrene

Benzpyrene is present in both coal tar (condensed vapors obtained upon heating coal) and tar from tobacco smoke. Both coal tar and tobacco-smoke tar cause cancer when painted on the skin of mice. Benzpyrene is not present in either tobacco or cigaret paper but is believed to be formed by the decomposition of cellulose when the cigaret is burned. As the chief constituent of the cell walls of plants, cellulose is present in both tobacco and paper. The molecular weight of cellulose is very large; it has been estimated to be between 50,000 and 500,000. The formula for cellulose may be given as $C_x(H_2O)_y$, where x and y are very large numbers.

If each carbon atom in a molecule forms two carbon-carbon bonds, the tetravalence of carbon allows each to form two bonds with other atoms. When hydrogen is the only element other than carbon present, the compound is called a *hydrocarbon*. We have drawn structural formulas for 12 hydrocarbons, namely, methane, ethane, propane, hexane, cyclopentane, cyclohexane, ethene, acetylene, benzene, toluene, naphthalene, and benzpyrene. If the element oxygen is present in addition to hydrogen and carbon, the compound may be:

An alcohol — Methyl alcohol, Ethyl alcohol

An acid — Acetic acid, Propanoic acid

An ether — Dimethyl ether, Methyl ethyl ether

ISOMERISM

A sugar

Glucose

or one of several other classes of compounds. When we consider that in addition to hydrogen and oxygen, the elements sulfur, phosphorus, nitrogen, fluorine, chlorine, bromine, and iodine are all frequently found in organic molecules, we may begin really to appreciate the variations possible.

Isomerism

We have thus far discussed four factors partially responsible for the multiplicity of carbon compounds: (1) the formation of carbon chains of various lengths, (2) the formation of carbon rings of various sizes, (3) the presence of multiple bonds, and (4) the presence of other elements. A final, important factor remains, the phenomenon of *isomerism*.

Look again at the structural formulas for the two compounds ethyl alcohol and dimethyl ether. What are their molecular formulas? It is apparent that they have the *same* molecular formula, C_2H_6O, yet they are very different compounds, one being an alcohol and the other an ether. The difference is emphasized by the fact that ethyl alcohol is a liquid at room temperature, while dimethyl ether is a gas! *Isomerism* is the existence of two or more *different* compounds with the same molecular formula. Compounds exhibiting this relationship to each other are called *isomers*. These particular compounds are called *structural isomers* because they have the same molecular formula but different structural formulas.

If we had continued drawing structural formulas for members of the hydrocarbon family consisting of methane, ethane, propane, and so on, we would have encountered this same phenomenon with the next molecular formula, C_4H_{10}, for the butanes, the four-carbon hydrocarbons of this type. There are two different structures, both of which correspond to the molecular formula of C_4H_{10}:

Normal butane   Isobutane

Another kind of isomerism among carbon compounds is called *stereoisomerism*. Stereoisomers are compounds with the same molecular formula and the same structural formula but with different spatial arrangements of the atoms. Stereoisomerism can be further subdivided into two types, *geometric isomerism* and *optical isomerism*. We shall consider simple examples of each type in order to point out the relationships involved.

A special property of the double bond, namely, restricted rotation, is responsible for geometric isomerism among carbon compounds. To explain how spatial relationships of atoms within a molecule may produce geometric isomerism, we shall discuss molecules of compounds containing carbon, hydrogen, and chlorine. The molecular formula $C_2H_4Cl_2$ represents two compounds which are structural isomers, since in one of the compounds both chlorine atoms are bonded to the same carbon atom and in the other they are attached to different atoms:

$$\begin{array}{cc} \text{Cl Cl} & \text{H Cl} \\ \text{H—C—C—H} & \text{H—C—C—Cl} \\ \text{H H} & \text{H H} \\ \text{(I)} & \text{(II)} \\ \text{1,2-Dichloroethane} & \text{1,1-Dichloroethane} \end{array}$$

Both compounds are dichloroethanes, the numbers in the names being necessary to give the location of the chlorine atoms. Consider the structural formula below. What is its relationship to the compounds 1,2-dichloroethane and 1,1-dichloroethane?

$$\begin{array}{c} \text{Cl H} \\ \text{H—C—C—Cl} \\ \text{H H} \\ \text{(III)} \end{array}$$

If one remembers that freedom of rotation is a fundamental property of the single bond, it will be clear that this formula is just another way of representing the molecule of 1,2-dichloroethane. It is exactly the same molecule as that represented by (I). In fact, the two structural isomers whose formulas are given in (I) and (II) are the *only* isomers of this molecular formula. All other variations are simply due to permitted rotations.

Now let us consider compounds with the molecular formula of $C_2H_2Cl_2$. There are two structural isomers corresponding to this formula.

$$\begin{array}{cc} \text{Cl} \quad \text{Cl} & \text{H} \quad \text{Cl} \\ \text{C=C} & \text{C=C} \\ \text{H} \quad \text{H} & \text{H} \quad \text{Cl} \\ \text{(IV)} & \text{(V)} \\ \text{1,2-Dichloroethene} & \text{1,1-Dichloroethene} \end{array}$$

ISOMERISM

Bearing in mind that rotation about the double bond is *not* permitted, what relationship does the following structural formula have to the two above?

$$\underset{H}{\overset{Cl}{>}}C=C\underset{Cl}{\overset{H}{<}}$$

(VI)

This is a different molecule. It has the same molecular formula as the compounds above and is not a structural isomer of 1,2-dichloroethene since it also has one chlorine and one hydrogen atom on each carbon atom. These two compounds [(IV) and (V)] are known as *geometric isomers*. The prefixes *cis-* and *trans-* are used to distinguish between molecules in which similar groups are on the same side of the double bond (*cis-*) and those in which they are on opposite sides (*trans-*):

cis-1,2-Dichloroethene   *trans*-1,2-Dichloroethene

Geometric isomerism will be found whenever the two carbon atoms of a double bond each have two different atoms or groups of atoms attached.

The second type of stereoisomerism, known as *optical isomerism*, is found only in three-dimensional molecules. To explain optical isomerism we shall consider a compound of the methane type, fluorochlorobromomethane, CHFClBr.

Fluorochlorobromomethane

Next, let us consider another arrangement of the attached atoms which corresponds to the mirror image of the first molecule.

(I)   (II)
      Mirror image

Although much more easily seen with models which can be manipulated, it should be clear from these diagrams that these two molecules are *different* compounds. No permitted manipulation of model (II)—whether rotation about the single bond or rotation of the whole molecule—can give an arrangement equivalent to (I). Suppose we hold the carbon and hydrogen atoms in model (II) in fixed positions and rotate the other three in a counterclockwise direction until the fluorine atom is in the same relative position as in (I). The result is that the bromine and chlorine atoms are now in noncorresponding positions. We say that the two forms are nonsuperimposable, which is merely another way of saying that they are *not* identical. *Optical isomerism* exists between two molecules if they have the same molecular formula and structural relationship and one is the mirror image of the other but nonsuperimposable.

This example was perhaps the simplest molecule one could find to show optical isomerism. In larger and more complex molecules instead of four atoms attached to the carbon one more frequently finds four *groups of atoms*. The possibility of optical isomerism arises whenever the carbon atom is tetrahedrally bonded to four different groups.

We have discussed at some length factors which contribute to the multiplicity of carbon compounds. The single most important factor, however, remains the extent to which the carbon atom has the ability to form homonuclear bonds.

This property is present to a surprisingly lesser degree in silicon, the element directly under carbon in the periodic table. Silicon has allotropic forms which correspond to diamond and graphite but is never found in nature in the free state. Compounds of hydrogen and silicon based upon homonuclear silicon-to-silicon bonds are also known, but their numbers and varieties are very limited. No silicon chains containing more than eight silicon atoms are known.

Silicon compounds, relatively unimportant in this classification, move to a position of first importance in the element-to-oxygen-to-element type of polymeric molecules and ions considered next.

Heteronuclear Bonding

Each of the nonmetals of the third period in the periodic table forms an anion with four oxygen atoms at the vertices of a regular tetrahedron and the nonmetal atom at the center. These ions have different charges, of course, because of the different numbers of valence electrons on the nonmetal atoms. The nonmetal-to-oxygen bond varies in polarity with the differences in electronegativities among the nonmetal atoms (Fig. 6-11).

As can be seen from a consideration of the respective electronegativities, the Si—O bond should be the most polar of the four shown, and the Cl—O bond should be the least polar. The electron pair of the bond

HETERONUCLEAR BONDING

Figure 6-11 Nonmetal-oxygen tetrahedra for third-period nonmetals.

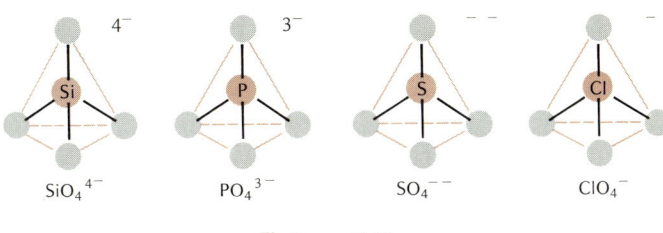

SiO$_4^{4-}$ PO$_4^{3-}$ SO$_4^{--}$ ClO$_4^-$

Electronegativities
Si = 1.8   P = 2.1   S = 2.5   Cl = 3.0   O = 3.5

is more equally shared in the Cl—O bond than in any of the others. The result is as might be predicted. Electrons of oxygen atoms on the perchlorate ion are almost *never* shared with atoms outside of the tetrahedron, so that the ion is hardly ever extended. Extension occurs somewhat more easily with the sulfate ion, considerably more easily with the phosphate ion, and readily and consistently and in all directions with the silicate ion.

CHLORINE

The only reasonably stable extended unit of the ClO$_4^-$ type is Cl$_2$O$_7$, an explosive oxide:

O—Cl—O—Cl—O (with O substituents)

Dichlorine heptoxide

SULFUR

The pyrosulfate ion, S$_2$O$_7^{--}$, similar in structure to the dichlorine heptoxide molecule, results from the sharing of one of the atoms of a sulfate tetrahedron with another:

[O—S—O—S—O]$^{--}$ (with O substituents)

The pyrosulfate ion

The molecular compound SO$_3$ has two forms (Fig. 6-12). One consists of rings of three SO$_4$ tetrahedra, and the other is a chainlike fibrous

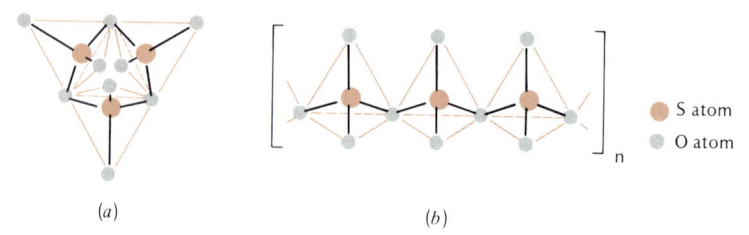

Figure 6-12 Molecular variations of SO_3: (a) ring and (b) chain forms.

structure. Both forms result from the sharing of *two* oxygen atoms of the SO_4 tetrahedron.

PHOSPHORUS

The phosphate ion can and does share none, one, two, three, or four of its oxygen atoms with other tetrahedra. This results in many varieties of one-, two-, and three-dimensional phosphate compounds. One large and important class, the *metaphosphoric acids*, corresponds to the empirical formula $(HPO_3)_n$, where n may be 3, 4, 5, 6, ... Some of these are ring molecules, as in $H_3P_3O_9$, $H_4P_4O_{12}$, and $H_6P_6O_{18}$, and others, in which the value of n is a very large number, consist of long branched-chain molecules (see Fig. 6-13). The long chains may be cross-linked, so that, depending upon the extent of the linkage, the acid may be a sticky, viscous liquid or actually set to a glassy mass.

SILICON

As the carbon-to-carbon bond is fundamental to the thousands of organic compounds, so the silicon-to-oxygen-to-silicon bond is fundamental to the many inorganic mineral compounds. Naturally occurring silicates include sand, granite, flint, agate, quartz, onyx; the gem stones emerald, amethyst, zircon, aquamarine, and opal; the water-softening zeolites; talc, clays, and the micas; and the fire-resistant fiber asbestos. Manufactured building materials such as bricks, mortar, and cement are silicates, as are ceramics and glass.

The oxygen atoms at the four corners of the SiO_4 tetrahedra have a great tendency to be shared with other silicon atoms. If none or one is shared, a discrete ion results. If two oxygen atoms per SiO_4 unit are shared, chains or rings are the result. If three oxygen atoms are shared, two-dimensional sheets are formed, and if all four oxygen atoms are shared, three-dimensional solids are the result. Let us consider these structures and some examples of each class.

HETERONUCLEAR BONDING

When no oxygen atom is shared, the ion is the *orthosilicate ion*, SiO_4^{4-}, and with one oxygen atom shared, the *pyrosilicate ion*, $Si_2O_7^{6-}$. Minerals containing these ions are zircon, $ZrSiO_4$, and thorveitite, $ScSi_2O_7$.

When two oxygen atoms are shared, the anion may be a ring structure as in beryl, $Be_3Al_2(Si_6O_{18})$ (Fig. 6-14). Sharing two oxygen atoms also gives rise to the fibrous minerals, the *pyroxenes* [Fig. 6-15(a)]. In the double-chained *amphiboles* [Fig. 6-15(b)], two and three oxygens are shared alternately. Most commercial asbestos is an amphibole.

Figure 6-13 Metaphosphoric acid molecules: (a) ring and (b) branched-chain molecules.

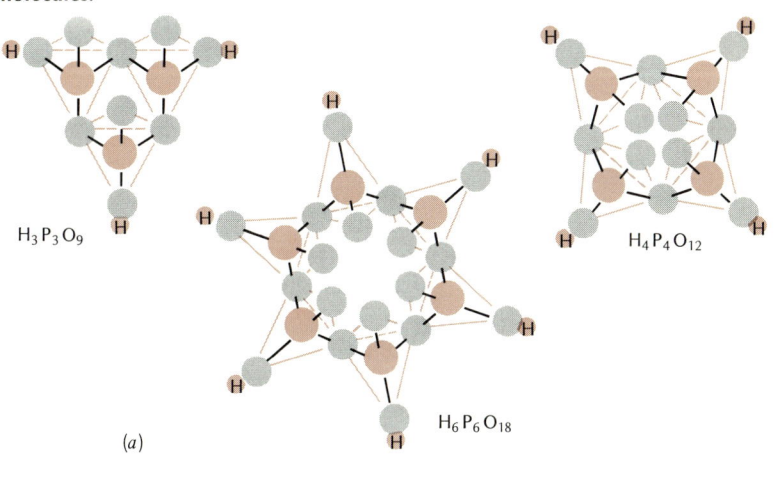

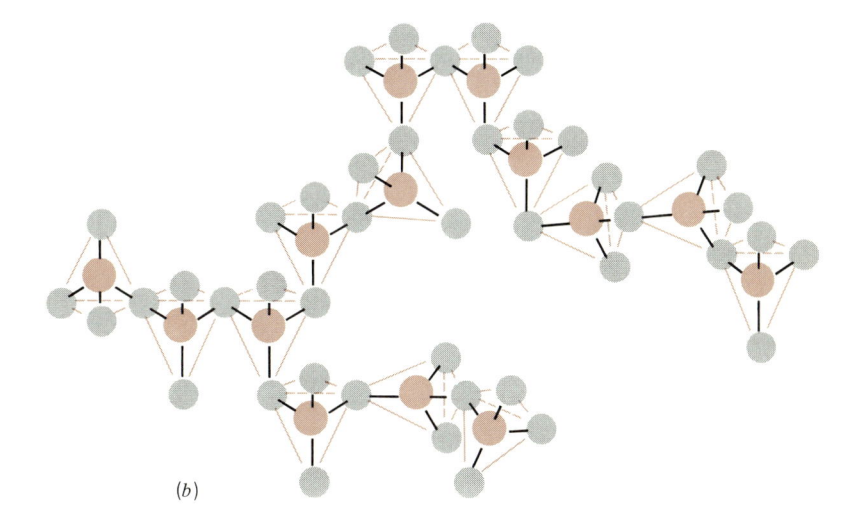

$H_3P_3O_9$

$H_6P_6O_{18}$

$H_4P_4O_{12}$

(a)

(b)

112 POLYMERIC MOLECULES AND IONS

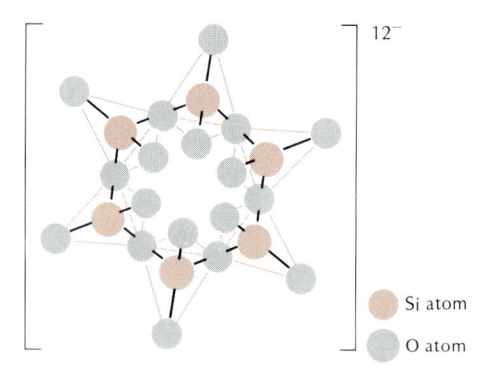

Figure 6-14 Anion of beryl.

Figure 6-15 Fibrous silicate minerals: (a) silicate (pyroxene) chain; (b) silicate (amphibole) chain.

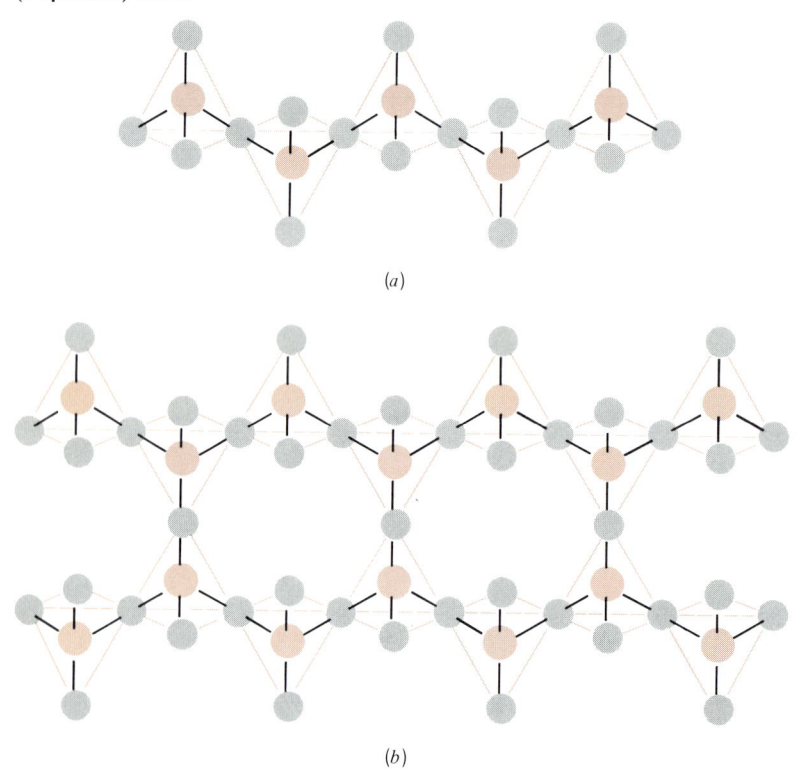

The sharing of three oxygen atoms by each SiO$_4$ tetrahedron gives two-dimensional sheets, a layered structure characteristic of talc, the micas, and clay minerals (see Fig. 6-16).

Sharing all four oxygen atoms results in many different three-dimensional minerals. If no element other than silicon and oxygen is present, pure silica quartz, SiO$_2$, is the result (see Fig. 6-17). Replacement of the silicon atoms by aluminum atoms in varying proportions is characteristic of the feldspars, one of the largest classes of naturally occurring minerals. Feldspar occurs in almost every major kind of rock.

Silicates make up over 25 percent of the known mineral species and over 90 percent of the earth's crust.

Recently, silicon-nitrogen-silicon polymers, which do not occur in nature, have been synthesized in the laboratory. The structures of these compounds can be represented by the following general formula, where n is a variable and the R's are carbon-hydrogen groups:

The R's may be different or all the same. The nature of these polymers varies with the size and nature of R, but in general they are heat-resistant; unlike the silicates, they are tough, flexible, and rubbery.

Figure 6-16 Sheetlike silicates.

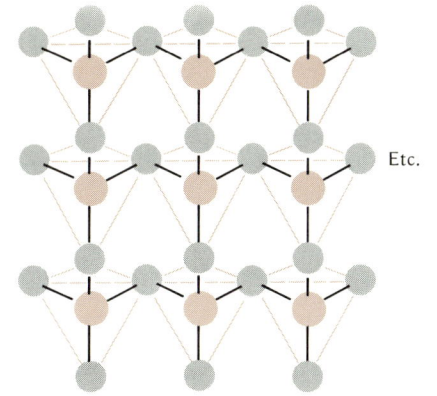

Figure 6-17 The structure of silica quartz, SiO$_2$.

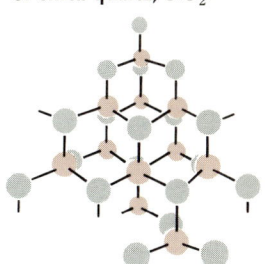

114 POLYMERIC MOLECULES AND IONS

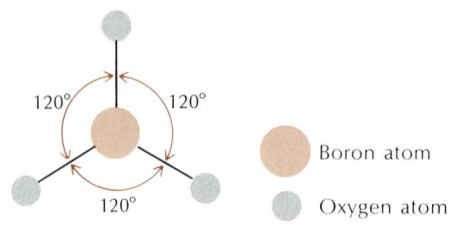

Figure 6-18 The structural unit of the borates.

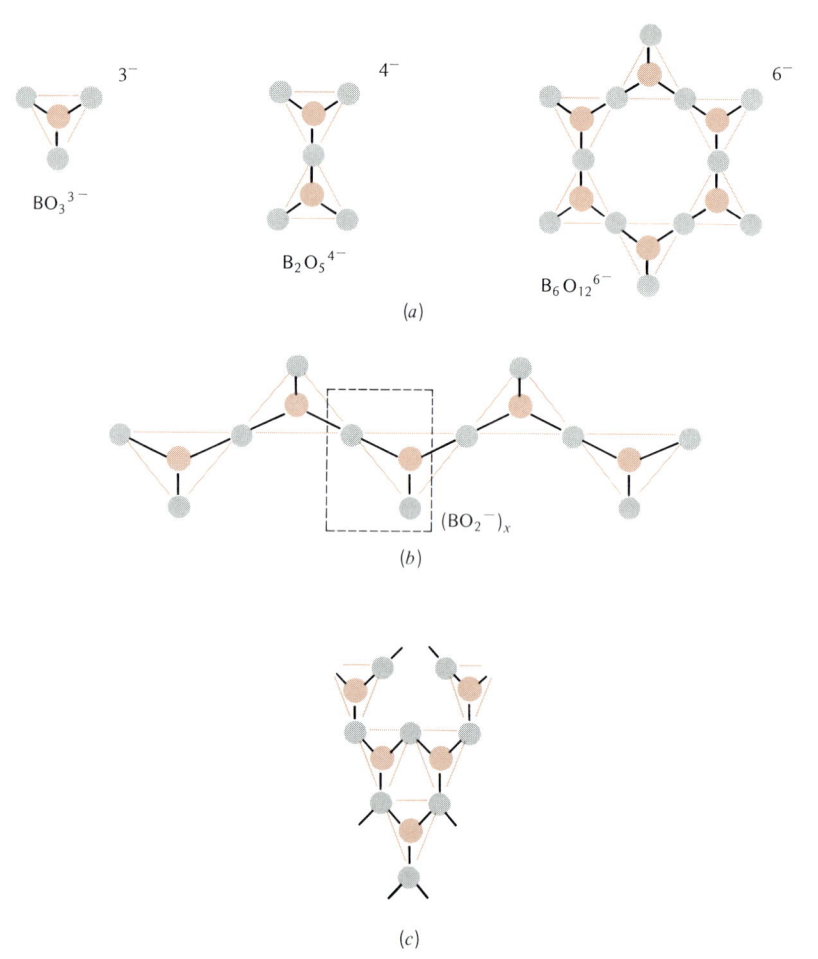

Figure 6-19 Borate structures: (a) discrete borate molecules; (b) borate chain molecules; (c) two-dimensional borate sheet.

BORON

The *borates*, boron-to-oxygen-to-boron bonded minerals, are similar in some respects to the silicates. Since the boron atom has only three valence electrons, the structural unit for these compounds is a two-dimensional triangle with the boron at the center and the three oxygen atoms at the vertices (Fig. 6-18). (The fact that boron does *not* complete its octet when bonding was discussed in Chap. 5.)

In a fashion similar to that for the silicates, if no oxygen atom is shared, the BO_3^{3-} ion results, and if one oxygen atom is shared, the $B_2O_5^{4-}$ ion. If two oxygen atoms are shared, chains or rings result, and if all three oxygen atoms are shared, two-dimensional sheet anions are formed. Because it is restricted to three bonds, and these in three planar-triangular bonding directions, there are no three-dimensional structures of the borates (see Fig. 6-19).

Questions

1 Distinguish between the following terms:
 a Polymerism, allotropism, and isomerism
 b Homonuclear bond and heteronuclear bond
 c Structural isomerism and stereoisomerism
2 What information is provided by each of the following types of formulas?
 a Empirical formula b Electron formula
 c Structural formula d Molecular formula
3 Give the molecular and electron formulas for the elemental gases nitrogen, oxygen, and chlorine.
4 What molecular formulas are used to represent solid sulfur and solid white phosphorus?
5 The chemical formula for carbon in the elemental state is generally given simply as C. Which of these two possible formulas would better represent carbon: C_∞ or C_n?
6 The S_2 molecule is paramagnetic. Suggest an electron formula for this molecule.
7 The large number of carbon compounds is chiefly attributable to what property of the carbon atom?
8 Draw electron formulas and structural formulas for the following molecules:
 a CCl_4 b NH_3 c H_2O
 d HI e SiF_4 f CH_3Cl
 g PCl_6^- h SO_2 i CS_2
 j BF_3 k Br_2 l CO
 m C_2H_2 n CO_2
9 What would you predict for the geometry of the acetylene molecule?

116 POLYMERIC MOLECULES AND IONS

10 Draw all possible structural isomers of C_5H_{12}.
11 How many structurally different monochloropentanes are there? Dichloropentanes?
12 The compound whose structural formula is

```
                    H
                    |
                  H—C—H
                    |
      H   H   H    |    H   H
      |   |   |    |    |   |
  H—C—C—C—C—C—C—H
      |   |   |    |    |   |
      H   |   H   H   H   H
          |
        H—C—H
          |
          H
```

has an optical isomer. Can you discover which carbon atom in the molecule has four *different* groups attached?

13 How many structural isomers of the double-bond-containing compound $C_2H_2Br_2$ are there? Do any of these structural isomers have geometric isomers? Draw structural formulas for all these forms.

14 Consider the following molecular formulas.
   a Which represent only one compound?
   b Which have structural isomers?
   c Which have optical isomers?
   d Which have geometric isomers? Write the structural formulas for all geometric and structural isomers.

   $C_2H_3Br_3$ CH_4 C_2H_3IBrCl
   $C_2H_2I_2$ C_3H_7Cl C_2H_6
   $C_2H_4Br_2$ C_2H_5I C_3H_8O

15 Draw the structural formulas for the following extended (polymeric) molecules and ions.
   a $H_3P_3O_9$ b $H_4P_2O_7$ c $H_4P_4O_{12}$
   d $H_6P_6O_{18}$ e $Si_3O_9^{6-}$ f $Si_2O_7^{6-}$

16 Place the five terms isomerism, geometrical isomerism, optical isomerism, stereoisomerism, structural isomerism, in the correct boxes to demonstrate the proper set-subset relationships among them.

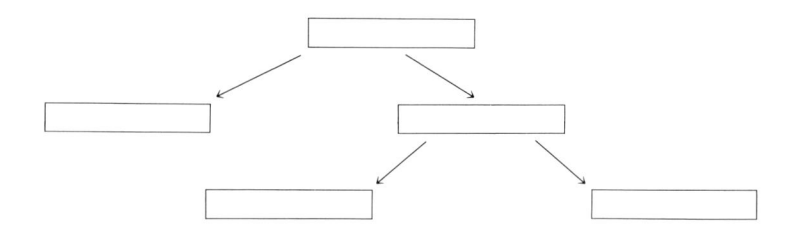

QUESTIONS

17. Are polymeric ions and molecules formed only through homonuclear bonding?
18. Would you predict greater or lesser tendency to form extended ions involving bromine-to-oxygen bonds than chlorine-to-oxygen bonds? With bromine-to-oxygen bonds than with selenium-to-oxygen bonds?
19. What molecular formulas might be used to distinguish between the two forms of SO_3?
20. What accounts for the fact that there are three-dimensional silicates but no three-dimensional borates?

CHAPTER 7   THE PHYSICAL STATES OF MATTER: FLUIDS

As easily as on land, the water strider moves over the surface of the water. Its security is due to the phenomenon known as *surface tension*. Intermolecular forces in liquids are manifested in the formation of a surface "film," which the legs of the insect depress but do not break. (Dr. Syd Radinovsky, Millersville State College, Millersville, Pa.)

CHAPTER 7   THE PHYSICAL STATES OF MATTER: FLUIDS

In Chap. 1 a consideration of the classifications of substances possible through observing, classifying, and measuring the obvious, outward properties of matter suggested the three physical states of solid, liquid, and gas as useful classifications. One of the questions we have proposed to answer is why one substance is a gas at room temperature while others are liquids or solids. In previous chapters we have considered the forces responsible for the formation of molecules and ions. In this chapter we turn our attention to forces between molecules and ions. These forces account for the differences between the three physical states.

It is apparent that different properties are useful in describing substances in different states. For instance, specific solid substances are characterized by a melting point, and specific liquids by a boiling point. When we wish to describe not a substance but a state—solid, liquid, or gas—we must similarly look for properties which all substances *in that state* exhibit and which distinguish that state from the others. For liquids and solids, ability to reach a boiling point and a melting point, respectively, are such characteristics. On the other hand, such qualities as color and odor are not useful to distinguish any state, since all states may exhibit color or odor *or* have no color or odor.

Density provides a useful clue in distinguishing the gaseous state if we keep certain qualifications in mind. Densities of gases at standard conditions of temperature and pressure (conditions we shall subsequently define) vary from less than 0.0001 to 0.0098 g/cm^3, and densities of liquids vary from 0.6 to 13.5 g/cm^3. These figures might lead one to suppose that the physical states can be differentiated on the basis of density, but the densities of solids lie in a range very similar to that of liquids: from 0.3 to 21 g/cm^3. The comparative densities of a single substance in the three states, however, are differentiating (see Table 7-1). From the figures in this table we might safely conclude that for a given substance, the solid and liquid states are many times, often several hundred times, as dense as the gaseous state.

Two other properties, related to each other, differentiate the vapor state from the liquid or solid, namely, *rapidity of diffusion* and *ease of compression*. A quantity of a gas placed in a container quickly diffuses to all parts of the container. We say that the gas "assumes the shape and volume of the container." Furthermore, the gas behaves this way whether it is in an evacuated container or one already containing another gas. The presence of another gas and thus the pressure in the container will be observed to retard the rate of diffusion, but, given time, diffusion will take place to the same degree as in an evacuated container. If the container is a cylinder with a movable piston for a top (Fig. 7-1), the gas can be compressed; and a large compression is possible with a relatively small expenditure of energy. Liquids and solids can hardly be compressed at all, even with great pressures.

THE PHYSICAL STATES OF MATTER: FLUIDS

TABLE 7-1 Comparative Densities of Substances in Three Physical States

Substance	Density of Gas at Boiling Point, g/cm³	Density of Liquid at Melting Point, g/cm³	Density of Solid, g/cm³
Ar	0.00179	1.40	1.65
Cl_2	0.00312	1.56	2.2
CO_2	0.00196	1.10	1.56
F_2	0.00170	1.11	1.5
H_2O	0.000804	1.00	0.917

The most obviously characteristic property of liquids is their ability to *flow*. A quantity of a liquid placed in a container flows to assume the shape of the container, although its volume remains independent of the container. It will fill the container from the bottom up and maintain a liquid *surface* if the container is not filled. However, the ability of a gas to diffuse and of a liquid to flow may be considered the same property, exhibited to different degrees. For this reason these two states are frequently given the single classification of *fluids*. Other evidence lends validity to the consideration of the liquid state as an extension of the gas state. Above a certain temperature, characteristic for each liquid, the surface between the liquid phase and the gas phase disappears and the substance exists in one phase only.

Gases diffuse to assume the shape of the container. Liquids flow to assume the shape of the container. Solids maintain their shape inde-

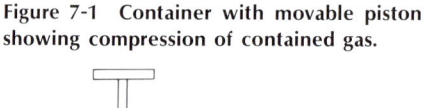

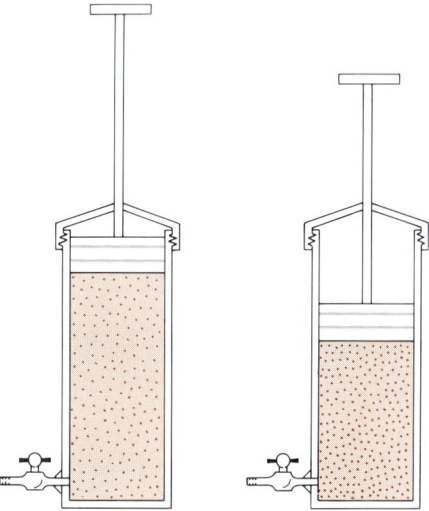

Figure 7-1 Container with movable piston showing compression of contained gas.

pendent of the container. Furthermore, a close inspection of solid substances shows them to have, in addition to a definite volume and a definite shape, a geometric structure which is manifested in their characteristic crystalline regularity. We defer consideration of the solid state to the next chapter, and in this chapter seek an explanation for the properties of fluids.

Properties of Gases and the Gas Laws

The ability of a gas to be compressed is characteristic. A careful consideration of this property requires an understanding of the concept of pressure and of methods for measuring *pressure*. Popular notions of pressure frequently confuse it with *force*. In ordinary terms, force is the total push, while pressure is the push on a unit area. Compare the relative effectiveness of a hammerblow in driving a blunt metal rod or a nail into the ground. In both cases the total force of the blow is the same, but by reducing the area on which it is applied in the case of the nail, pressure is multiplied several times. Air pressure, or *atmospheric pressure*, is approximately $14\frac{1}{2}$ lb/in.^{2}. The weight of the atmosphere is distributed over every surface so that the force on each square inch of surface area is approximately $14\frac{1}{2}$ lb.

Atmospheric pressure is measured by an instrument called a *barometer*, which consists of a closed tube filled with mercury and inverted in a pool of mercury (see Fig. 7-2). Pressure is expressed by a barometer not in terms

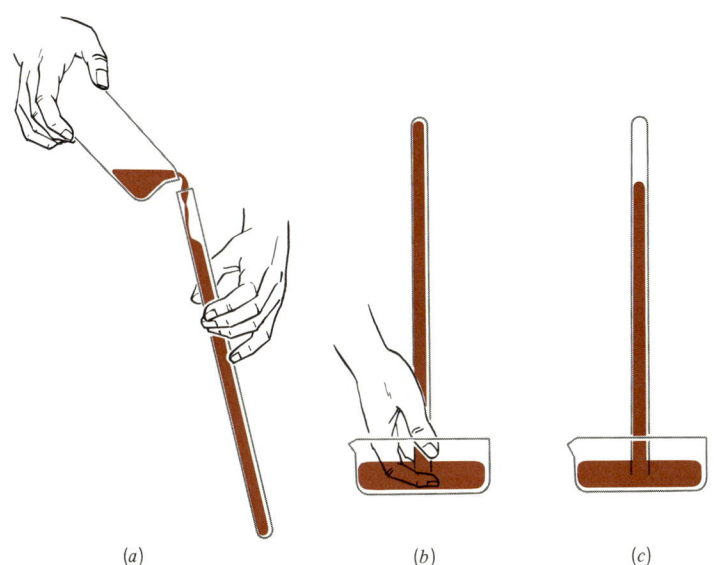

Figure 7-2 Constructing a barometer.

(a)   (b)   (c)

124 THE PHYSICAL STATES OF MATTER: FLUIDS

of force per unit area but in terms of the height of a column of mercury which will be supported by the air pressure on the surface of the pool of mercury. Any liquid, of course, can be used in a barometer, and early ones utilized water. In Fig. 7-3 are shown the comparative heights of several liquids which would be supported by the pressure of the atmosphere on a day when the mercury barometer gives a reading of 760 mm. The great advantage of mercury as the liquid in the barometer should be obvious. If the concept of pressure is understood, the student will also accept the fact that the diameter of the barometer tube has no effect upon the height (see Fig. 7-4).

While it may be common knowledge that increasing the pressure on a confined gas at constant temperature will reduce the volume of that gas, not so common is the understanding of the mathematical relationship between the volume and the pressure. It is a very simple one: the volume change is inversely proportional to the pressure change. If the pressure is doubled, the volume will be cut in half; if the pressure is tripled, the volume will shrink to one-third of its original value; if the pressure is halved, the volume will double, and so on. Stated formally: *the volume of a given sample of a gas is inversely proportional to the pressure at constant temperature.* This is called *Boyle's law* for Robert Boyle (1627–1691), the British chemist who first observed the relationship. It will be helpful here to define the conditions of temperature and pressure which scientists refer to as *standard conditions*. These reference conditions are 0.0°C and 1 atm, or 760 mm of mercury pressure, and are abbreviated STP for standard temperature and pressure.

John Dalton, whose influence upon atomic theory we have already noted (page 19), added much to the understanding of the behavior of gases. He found that when two gases from different containers at equal

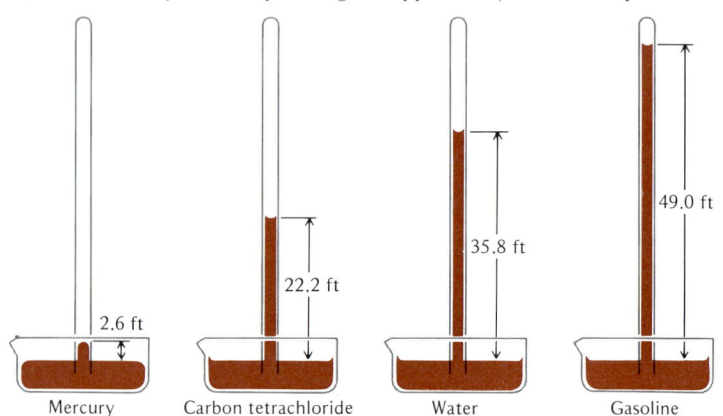

Figure 7-3 Comparative liquid heights supported by 1.00 atm of pressure.

Mercury	Carbon tetrachloride	Water	Gasoline
2.6 ft	22.2 ft	35.8 ft	49.0 ft

PROPERTIES OF GASES AND THE GAS LAWS

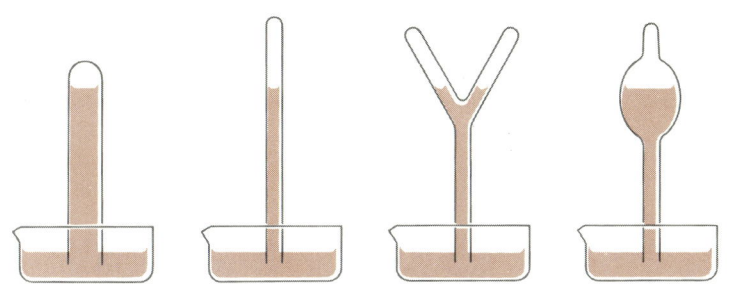

Figure 7-4 Neither cross-sectional area nor shape of tube affects the height of liquid supported by atmospheric pressure.

pressures and temperatures are allowed to mix, neither container shows any change in temperature or pressure. He found this to be true whether the two samples are the same or different gases—provided, of course, that they do not react chemically. Consider Fig. 7-5, which shows two adjacent 1-liter containers, one containing a sample of a gas at 1 atm pressure and the other an evacuated container. When the gas sample is allowed to diffuse into the empty container, what will be the resulting pressure on the whole system? Boyle's law tells us that doubling the volume will halve the pressure, so that the new pressure—on both parts of the container—will be $\frac{1}{2}$ atm. Now consider Fig. 7-6, again two adjacent 1-liter containers, each this time with a sample of a gas at 1 atm pressure. When these two are allowed to diffuse into each other, the pressure on the combined containers will remain the same because *each gas exerts the pressure it would exert if it occupied the container alone, and the total pressure is equal to the*

Figure 7-5 Two adjacent 1-liter containers. (*a*) Container at left is filled with a gas at 1 atm pressure; container at right is evacuated. (*b*) Partition is removed; diffusion has taken place.

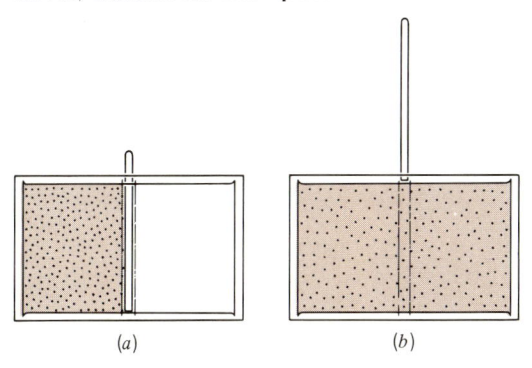

(*a*) (*b*)

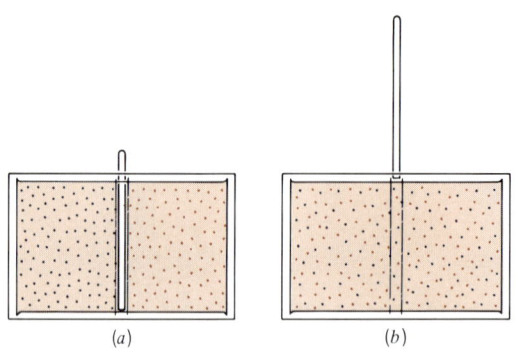

Figure 7-6 Two adjacent 1-liter containers. (a) Each container holds a sample of a gas at 1 atm pressure. (b) Partition is removed; diffusion has taken place.

sum of the partial pressures. This is a statement of *Dalton's law of partial pressures.*

At first glance, the relationship between the volume and the temperature of a gas at constant pressure may seem as simple as the volume-pressure relationship. If the temperature of a gas at constant pressure is increased, the gas will expand, and if the temperature is decreased, the volume will decrease. But the mathematical relationship is not as simple as might be expected, for if the temperature of a given sample of a gas at constant pressure is doubled from 20 to 40°C, the volume will be multiplied not by a factor of 2 but by a factor of $\frac{313}{293}$. If the same sample at 100°C has its temperature doubled to 200°C, the volume will be multiplied by a factor of $\frac{473}{373}$.

A good way to obtain a picture of a numerical relationship such as this one, and thus perhaps a clearer understanding, is to construct a graph. Measured at 0.0°C and 1 atm pressure 1 liter of a gas will occupy approximately the volumes given in Fig. 7-7 if the temperature is changed to that specified while the pressure remains constant. Figure 7-7 also shows the graph of these data.

We can use this graph to *interpolate,* or approximate, values intermediate to those given in the data table, and we shall also find that *extrapolation,* or extending the line of the graph, yields an interesting value. The extrapolated line (the dotted line in Fig. 7-7) crosses the temperature axis at approximately −273°C. At this temperature the corresponding volume reading is 0. Does this mean that continued cooling will somewhere in the vicinity of −273°C cause the gas sample to disappear? We hasten to assure the student that it does not. For one thing, extrapolation is a risky procedure. If we obtained experimental volume-temperature data for the low-temperature range or volume-pressure data for the high-

PROPERTIES OF GASES AND THE GAS LAWS

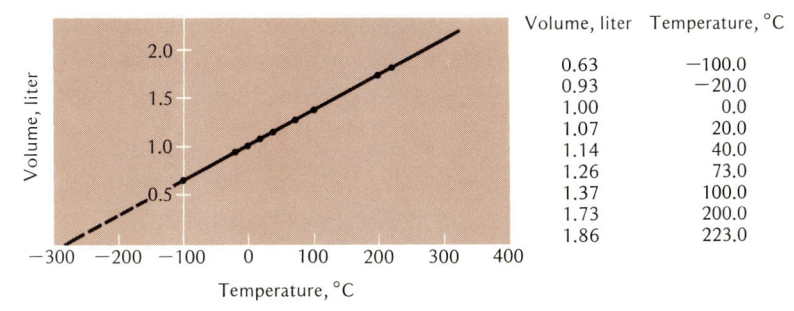

Volume, liter	Temperature, °C
0.63	−100.0
0.93	−20.0
1.00	0.0
1.07	20.0
1.14	40.0
1.26	73.0
1.37	100.0
1.73	200.0
1.86	223.0

Figure 7-7 Plot of temperature-volume data.

pressure range, we should find that the simple mathematical relationships and their straight-line representations no longer exist. These deviations are among the observed facts which any theory proposed as an explanation for the behavior of gases must also account for. Another fact that invalidates our extrapolation is that the relationship being considered is for volume and temperature of gases, only, and all gases liquefy or solidify before a temperature of −273°C is reached. The temperature of −273°C does have significance, however. It is called *absolute zero*, and scientists find a temperature scale with the same degree size as the Celsius scale but beginning at absolute zero particularly useful. This scale is called the *absolute* or *Kelvin scale* after Lord Kelvin (1824–1907), the British scientist who proposed it in 1848. Celsius temperatures of −100, 20, and 100°C, correspond to 173, 293, and 373°K. If we express the temperatures of Fig. 7-7 on the Kelvin scale (Table 7-2), we shall find that the mathematical volume-temperature relationship for gases is, after all, a simple one. If the Kelvin temperature of a gas at constant pressure is doubled, the volume will double. If the Kelvin temperature is halved, the volume will shrink

TABLE 7-2 Volume of a Sample of a Gas at Various Kelvin Temperatures

Volume, liters	Temperature, °K
0.63	173
0.93	253
1.00	273
1.07	293
1.14	313
1.26	346
1.37	373
1.73	473
1.86	496

to one-half of its original value. In other words, *the volume of a given sample of a gas at constant pressure is directly proportional to the Kelvin temperature.* This law is frequently referred to as *Charles' law,* for Jacques Charles (1746–1823).

Suppose the volume of a given sample of gas is held constant while the temperature is increased. We would predict that the pressure will increase. Experimental data show that *given a constant volume, the pressure of a gas is directly proportional to the Kelvin temperature.*

We have examined the relationships existing among the three variables of volume, pressure, and temperature for a gas, relationships collectively referred to as *the gas laws*. However, only one sample of a gas was considered in all cases except Dalton's law. Suppose we now consider cases in which the quantity of the sample is allowed to vary.

Given a rigid container (constant volume) and a constant temperature, what will be the effect upon the pressure of doubling the quantity of the gas within? Of reducing the quantity by half? (For the time being we shall take a change in quantity to mean only more or less of the original gas.) What seems intuitively probable is actually true and experimentally verifiable. *At constant temperature and volume, the pressure of a gas is directly proportional to the quantity.*

In a similar fashion, *at constant temperature and pressure, the volume of a gas is directly proportional to the quantity.*

The last relationship remaining to be considered, that between quantity and temperature, is an inverse one. For example, if a given sample of a gas is to show neither volume change nor pressure change when the quantity is increased, the temperature will have to be lowered. Stated formally: *the quantity of a gas at constant volume and pressure is inversely proportional to the Kelvin temperature.*

Let us summarize the properties of the gaseous or vapor state of matter which have been discussed in the preceding paragraphs. The *density* of the vapor state of a substance is many times less than that of the liquid or solid states. Gases *diffuse rapidly* and are *easily compressed. Simple mathematical relationships* exist among the variables of volume, temperature, pressure, and quantity of a gas, but *deviations* from these relationships are observed under very low temperatures or very high pressures.

Properties of Liquids

Let us next consider properties of liquids. We shall be able to list several properties common to all substances in the liquid state, but we shall not find simple mathematical relationships obtaining among them. Already noted are the facts that liquids *flow* to assume the shape of the container and that all liquids have characteristic *boiling points.*

All liquids can undergo both *evaporation* and *boiling*. The two phenomena yield the same end result, namely, the transition of a substance from the liquid to the gas state. Aside from this, however, differences between the two are more striking than similarities. Evaporation takes place at any temperature, although an increase in temperature will be found to increase the *rate* of evaporation. Boiling takes place at a given pressure at a characteristic constant temperature for each liquid. Variations in pressure cause variations in the boiling point: water boils on top of a mountain at a much lower temperature than at sea level. Evaporation is observed to be a surface phenomenon, because the rate of evaporation is increased by exposing a greater surface area of the liquid, but boiling appears to be a whole-body phenomenon, with the boiling temperature at constant pressure being essentially independent of the surface area. Careful observation shows that a liquid becomes cooler as it evaporates while a boiling liquid maintains a constant temperature throughout the boiling process.

Liquids exhibit *surface tension*. Objects which are heavier than water such as a needle or a razor blade can be floated on water with care, but if the surface "skin" is ruptured, they promptly sink to the bottom. The spherical form which freely falling drops of liquid tend to form is also due to the constrictive effect of surface tension.

All liquids can be caused to freeze, or solidify, and the *freezing point* (melting point) is characteristic for each liquid at a given pressure.

Liquids *cannot be compressed* appreciably but do undergo *thermal expansion*. Alcohol or mercury thermometers depend upon this property.

The properties of liquids which we have observed and hope to explain are the ability of liquids to *flow* and to undergo *thermal expansion;* the phenomena of *boiling, evaporation,* and *freezing;* the property of *surface tension;* and the fact that liquids *cannot be compressed* appreciably.

The Kinetic Molecular Theory

We now turn to the submicroscopic structure of matter in search of an explanation for these properties. Which of the common substances whose submicroscopic structure we have considered in detail are gases at room temperature? Which are liquids and solids? Among the elements discussed, all metals except mercury are solids, and all nonmetals are solids except bromine, which is a liquid, and the following which are gases: hydrogen, nitrogen, oxygen, fluorine, and chlorine. All the rare gases are gases at room temperature. Compounds whose structures have been discussed or for which electron formulas have been drawn and which are gases at room temperature include HCl, NH_3, CH_4, CO_2, CO, SO_2, C_2H_6, and several others. Liquids include H_2O, SO_3, CCl_4, CH_3OH, and others. Solids are more numerous: NaCl and other metallic chlorides, CaO and other metallic oxides, K_2S and other metallic sulfides, the sulfates, sulfites, nitrates, nitrites, carbonates, ammonium compounds, and silicates. Considering these substances as grouped here should make it possible for the student to answer the following question: *What submicroscopic structural characteristic do substances which exist as gases or liquids at room temperature have in common?* The answer, of course, is that the *substances which are fluids at room temperature have an essentially molecular character.* This is important. The molecular nature of fluids is the first assumption of the theory which best explains the properties of the various states of matter, the kinetic molecular theory.

Before we can consider this theory in detail, we must return to a consideration of the various forms of energy. In Chap. 1 we defined energy as the capacity to do work and discussed one form, *heat energy*, and its effects. Electricity, discussed in Chap. 2, is another form of energy, often called *electric energy*. We now must consider a third form of energy, namely, *kinetic energy*, or energy of motion, and at the same time its opposite, *potential energy*. These two forms are perhaps most clearly illustrated by a hammer precariously perched on the edge of a shelf. Because of its position, the hammer has a capacity to do work, which will consist of making a dent in the floor and a loud noise! As the hammer drops from the shelf, its energy is converted from energy of position to energy of motion, or from potential energy to kinetic energy. The kinetic energy given to a golf ball does the work not only of carrying the golf ball over a long distance but may also break a window. The amount of kinetic energy an object has depends both upon its mass and the speed of its motion. A bowling ball traveling at the speed of a golf ball through the air would accomplish much more work! The rifle bullet is smaller and lighter than the golf ball: its high speed enables it to accomplish more.

We are now ready to consider the basic assumptions of the kinetic molecular theory. This theory, as applied to gases, consists of two major assumptions. The first describes the *molecular nature* of all gases, and the second deals with the *kinetic energy* of the gas molecules.

The theory assumes that all gases consist of molecules. Since it has been observed that the physical behavior of all gases under ordinary conditions is identical, it follows that properties of the gas molecules themselves, such as size or polarity, have little if any effect upon physical behavior. The observation of the complete and unrestricted diffusion of gases in any container is taken to imply that the molecules of a gas are in constant, rapid, random, straight-line motion until colliding with each other or with the sides of the container. It must be assumed that there is no net loss of energy (although there may be a transfer) in these collisions, or the gas would "collapse" to the bottom of the container. Collisions of this nature are called *elastic*.

Stated concisely, then, the first assumption is this: *All gases consist of minute particles called molecules. The size of these molecules is negligible in comparison with the distances between them, and there are no attractions*

THE KINETIC MOLECULAR THEORY

between the molecules. The molecules are in constant, rapid, random, straight-line motion until colliding with the walls of the container or with each other, and all such collisions are perfectly elastic.

Second, the theory assumes that the velocity of the straight-line motion of the molecules is dependent upon the temperature. If a gas is heated, the velocity increases; if a gas is cooled, the velocity of the molecules is reduced. On an everyday level, this might be described as the conversion of heat energy to kinetic energy. On the molecular level it is more easily seen that the two "forms" of energy are really one and the same. Heat *is* the molecular motion, and temperature is a measure of the vigor of that motion. Furthermore, it is assumed that at the same temperature, the average kinetic energy of the molecules of all gases is the same. We have seen earlier that kinetic energy depends upon both the mass and the velocity of the moving particle. It is apparent, then, that in this second part of the theory we cannot disregard differences between molecules altogether. (Note that in the statement of the first assumption no mention was made of the comparative *masses* of gas molecules.) If two particles in motion possess the same kinetic energy while having different masses, it follows that their velocities must also be different. The heavier particle must be traveling more slowly and the lighter particle more rapidly in order that the *effect* of the motion of each is the same. (The golf ball would require a high speed to have the same effect as the bowling ball at a slow speed.) Thus the molecule of hydrogen, the lightest of all gases, has a velocity at any temperature many times that of any other gas and diffuses the most rapidly, while heavier, slower-moving gases diffuse more slowly. A concise statement of this second assumption of the kinetic molecular theory is this: *The velocity of the molecules of a gas is dependent upon the temperature, and at the same temperature the average kinetic energies of the molecules of all gases are the same.*

THE PROPERTIES OF GASES

The kinetic theory satisfactorily explains the gas-state properties of low density, ease of compression, and rapidity of diffusion. By disregarding the size of molecules and the attractions between them, we postulate a substance which behaves as if it consisted of points in space and which thus follows mathematical rules in its physical behavior. Pressure on the container walls is determined by the number of collisions per unit area of the wall per unit time. (Each collision may be thought of as a tiny "push.") If a sample of a gas is heated while the volume is held constant, the velocity of the molecules is increased, and so the number of collisions increases, which we measure by an increase in pressure. This same effect can be brought about by holding temperature constant while reducing the volume of the gas container—result: more collisions and higher pressure—or by introducing more gas into the container while holding volume and temperature constant—result: more collisions and higher pressure. With a little thought, the assumptions of the theory permit a complete explanation of the properties of gases which we have observed, except for the fact that at conditions of high pressures or low temperatures marked deviations from the gas laws occur. We can complete the explanation of gas properties by carefully stating that the assumptions are valid only at what have been described as "ordinary conditions" of temperature and pressure. What is it, then, that happens under extraordinary conditions? The assumption that the size of the molecules is negligible depends upon there being relatively large distances between them. At high pressures these distances are reduced to the point where the actual size of the molecules becomes important when total volume is considered. Thus at high pressures, a pressure increase causes less decrease in volume than would be predicted by a simple application of Boyle's law. The assumption that molecules have no attraction for each other depends upon the large distances between them and their high velocities. At very low temperatures these attractions for each other, while not increased, become more important as the molecules move more slowly and get closer together. So it is that at low temperatures the volume of a gas is reduced by a greater factor than Charles' law would predict from the temperature reduction. Even these deviations, then, rather than contradicting the theory simply give us a clearer understanding of it and of why it works so well under ordinary conditions.

AVOGADRO'S HYPOTHESIS AND THE MOLE

We have referred to the quantity of a gas without indicating the units in which that quantity might be measured. Under ordinary conditions, for volume, temperature, and pressure relationships, the molecules of a gas may be considered to be points in space with no properties. Quantities, then, can be measured only in terms of *numbers* of molecules. Furthermore, if all gases are to be considered this way, it follows that *equal volumes of all gases at the same conditions of temperature and pressure contain the same number of molecules*. This statement, called *Avogadro's hypothesis*, for Amedeo Avogadro (1776-1856), who suggested it in 1811, may seem a natural and logical consequence of the kinetic theory, but it took almost 50 years for this hypothesis to gain acceptance!

In some relationships involving gases the mass of the molecule cannot be disregarded. We have seen one example of this concerning the molecular velocity and resulting rate of diffusion. For this, and for other situations, we need to be able to describe the quantity of a gas in terms of its identity as well as the number of molecules. The mass of a single molecule is far too small a quantity to be useful. The *molecular weight* of a substance is a useful quantity unit. It is a relative weight, as is the

133 THE KINETIC MOLECULAR THEORY

TABLE 7-3 Gram Molecular Weights of Several Substances

Substance	Grams
Hydrogen, H_2	2.0
Oxygen, O_2	32
Sulfur dioxide, SO_2	64
Methane, CH_4	16

atomic weight (see Chap. 3). The molecular weight of a substance is the sum of the respective atomic weights of the atoms which compose it, each weight taken the number of times indicated by the molecular formula. The atomic weight of carbon is 12 and of hydrogen is 1.0. It follows, then, that the molecular weight of methane, CH_4, is 16. The molecular weight of SO_2 is 64, and of hydrogen gas, H_2, is 2.0.

Consider 1 liter of hydrogen gas, and 1 liter of methane gas at the same conditions. Avogadro's hypothesis tells us that each liter will contain exactly the same number of molecules. We also know that the liter of methane will weigh exactly 8 times as much as the liter of hydrogen, by comparing molecular weights. The converse of this is also true, namely, any time two samples of H_2 and CH_4 are found to have weights in the ratio of 1:8, these samples contain the same number of molecules. To make use of this fact a new unit is defined. The *gram molecular weight* of a substance is that amount whose weight in grams is numerically equal to the relative molecular weight. The gram molecular weights of some common substances are shown in Table 7-3.

We know that these respective weights of these gases will contain the same number of molecules. This number has been determined to a high degree of accuracy and is called *Avogadro's number*. Its value is 6.02×10^{23}. Thus 2.0 g of H_2 contains 6.02×10^{23} molecules; 32 g of oxygen contains 6.02×10^{23} molecules; 1.0 g of hydrogen contains 3.01×10^{23} molecules, and so on. Avogadro's number of units is commonly referred to as a *mole* of the substance. In everyday life, the unit dozen parallels the mole. There are 12 oranges in a dozen, 12 apples, 12 eggs, etc. In each case the *weight* of the dozen is different.

The gram molecular weight or mole of any gas contains 6.02×10^{23} molecules. At STP this mole will occupy 22.4 liters, which is known as the *gram molecular volume*.

PROPERTIES OF LIQUIDS

We shall find that the observed properties of liquids can be explained by simple extensions of the kinetic molecular theory. The key to the whole explanation lies in the phenomenon of *condensation*. This transition from a gas to a liquid can be brought about by cooling the gas and by increasing the pressure upon it. In terms of the kinetic molecular theory, what happens at this point? As the temperature is lowered, the average kinetic energy of the molecules of the gas is lowered, and as pressure is increased, the molecules are forced nearer each other. Finally the point is reached at which the average kinetic energy of the molecules is not sufficient to overcome the intermolecular attractions, and we say that the gas has *liquefied*. Molecules are still free to move over and around and past each other, which explains the ability of a liquid to flow, but they do not have sufficient energy to separate from each other entirely. We may look ahead to the discussions of the next chapter and see that by a further lowering of the temperature the average kinetic energy (and thus the molecular motion) will be reduced still more, until the molecules become completely localized with respect to each other and we say that the liquid has solidified or frozen.

This explanation should also make it clear why neither liquids nor solids are appreciably compressible. The molecules are in as close proximity as the repulsive forces between molecules will permit at a given temperature. Further changes can come only as a result of reduced molecular motion. Thermal expansion and contraction is a property of both liquids and gases. Changes in the kinetic energy of the molecules result in changes in the intermolecular distances and thus in changes in volume.

Liquefaction takes place when the kinetic energy of the gas is sufficiently low for intermolecular attractions to become important. Thus for every gas there is a temperature above which it becomes impossible to cause liquefaction by pressure alone, because the kinetic energy is too high for the intermolecular attractions to restrict the movements of the molecules, no matter how small the distances between. This temperature is called the *critical temperature* and is characteristic for each gas.

In observation of liquids, the two phenomena of evaporation and boiling were noted and compared. Let us compare the two in terms of molecular action. When we speak of the kinetic energy of the molecules of a gas or of a liquid, it is possible only to speak of the *average* kinetic energy of the molecules. Because of collisions with each other and with the container, individual molecules at any instant may have a much higher or much lower kinetic energy than this average. If, in a liquid, a molecule *on the surface* momentarily attains a higher than average kinetic energy, it may be sufficient for it to escape from the attraction of the other molecules entirely. This escape of molecules into the atmosphere above the liquid is *evaporation*. The more surface exposed, the greater the chance for escape, and the higher the temperature of the liquid, the easier it becomes for the molecules to overcome intermolecular attractions. What is the effect upon the liquid which remains behind as the molecules with

THE KINETIC MOLECULAR THEORY

higher than average kinetic energy continue to leave? The average kinetic energy of those left behind becomes lower, and thus the temperature drops. In terms of molecular motion, then, we have a picture of the evaporation process and can see why it is favored by increased surface area and higher temperatures and why it results in a cooling of the liquid.

Suppose that instead of a liquid open to the atmosphere we cover an open container of liquid with a bell jar, as in Fig. 7-8. Now let us describe the molecular motions which take place. At first evaporation is the only process observed, as the higher-kinetic-energy molecules escape from the surface of the liquid. But in time, as the atmosphere above the liquid acquires more and more vapor molecules, some of these vapor molecules strike the surface of the liquid at an instant when their kinetic energy is low enough for them to be "captured" by the liquid intermolecular forces, and they remain. In fact, a point will be reached when the number of molecules escaping from the liquid at any instant exactly equals the number of molecules returning. Ordinary observation would tell us that evaporation has stopped. An understanding of the molecular behavior tells us that it has not stopped at all but is merely balanced by the rate of condensation. A situation of this type wherein two opposing processes are acting against each other to give the effect of no change is called an *equilibrium*. Where the processes are physical, like evaporation and condensation, it is called a **physical equilibrium**. This type of equilibrium is a dynamic state as opposed to a static state in which all change has ceased.

The tendency for a liquid to escape into the vapor state or to evaporate is a measurable property, called the *vapor pressure* of the liquid. In the

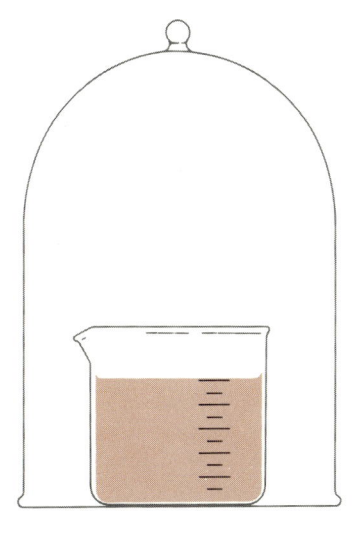

Figure 7-8 Open beaker of liquid under a bell jar.

example just discussed, the evaporation continued at a greater rate than the condensation as long as the vapor pressure of the liquid was greater than the vapor pressure of the gas molecules in the atmosphere above the liquid. At equilibrium the two vapor pressures are equal. We would say that the atmosphere had become *saturated*. This situation is encountered outside of the laboratory as well. If the amount of water vapor in the air is high enough—if it is humid enough—there will be no net evaporation. We need only think of the damp floors of shower rooms for an example.

How does the phenomenon of boiling differ from evaporation in terms of molecular motion? If a liquid which has reached an evaporation-condensation equilibrium with the atmosphere above it is heated, the average kinetic energy of the molecules of the liquid is increased and the tendency to escape, the vapor pressure, is also increased. The equilibrium is upset, and for a time evaporation will take place at a greater rate than condensation, until the atmosphere becomes saturated once more. Table 7-4 gives the vapor pressure for water at various temperatures.

The vapor pressure, described as a measure of the tendency to evaporate, can also be taken as a measure of the relative number of molecules with sufficiently high kinetic energy to have escaped into the vapor phase. If the figures in Table 7-4 are for water being heated under 1.00 atm, at 10°C only 9.2/760, or 1.2 percent, of the liquid molecules have sufficient kinetic energy to escape. At 20°C, 17.5/760, or 2.3 percent; at 50°C, 92.5/760, or 12.2 percent; and then finally, at 100°C, all the molecules have attained a high enough kinetic energy to vaporize and the water boils. Boiling is not a surface phenomenon, but a whole-body process; inhibited only by the requirements of transfer of heat through the body of the liquid, the entire liquid can enter the vapor state. The sizzle of a drop of water on a hot stove is literally an explosion of the entire drop into vapor form. While the liquid boils, it was noted that the temperature did not change. The heat that is still being supplied does not raise the kinetic

TABLE 7-4 Vapor Pressure of Water at Various Temperatures

Temperature, °C	Vapor Pressure, mm Hg
10	9.2
20	17.5
30	31.8
50	92.5
70	233.7
90	525.8
95	633.9
100	760.0

137 INTERMOLECULAR ATTRACTIONS

Figure 7-9 Schematic representation of the forces producing surface tension in liquids.

energy of the molecules any further but instead increases the potential energy of the molecules by allowing them to separate. The amount of heat which must be supplied to vaporize a gram of a substance at its boiling point is known as the *heat of vaporization*, and this amount gives some indication of the strength of intermolecular forces in a compound. The definition of the boiling point of a liquid is *that temperature at which the vapor pressure of the liquid equals the atmospheric pressure*. Boiling points given in tables are for an atmospheric pressure of 760 mm Hg.

An interesting application of Dalton's law of partial pressures can be made with these same vapor-pressure figures. Since this law tells us that the total pressure must be the sum of the partial pressures and that each gas exerts the pressure it would if it were in the container alone, the percent figures we derived also tell us what percent of the gas above the liquid is vapor and what percent is the original atmosphere (see Question 32).

Surface tension is a property of liquids by which they appear to maintain a "skin" at the surface. Consider the theoretical diagram of the molecules of a liquid in Fig. 7-9. The arrows are used to indicate the direction of intermolecular forces upon the molecules. A molecule at the center of the liquid is attracted equally in all directions, but a molecule at the surface is attracted only inward, toward the body of the liquid. This increased and directional attraction of and for the surface molecules results in curved liquid surfaces and spherical raindrops. It is also responsible for the fact that liquid surfaces behave differently than the interior of the liquid relative to floating objects.

Intermolecular Attractions

The explanation of the properties of gases in terms of the kinetic molecular theory was quite simple, since under ordinary conditions the physical behavior of all gases is the same. However, in addition to explaining the properties of the liquid state we must be able to account for differences in these properties, because the physical behavior of liquids shows great variations.

Liquid hydrogen boils at −252.7°C, liquid chlorine at −34.6°C, carbon disulfide at 46.3°C, and water at 100°C. These relative boiling points must indicate the relative strengths of the intermolecular forces which have to be overcome by the kinetic energy of the molecules before vaporization can take place. The nature of the intermolecular forces depends upon the size and the geometry of the molecules themselves. We shall describe three types of intermolecular forces. It should be remembered, however, that these are descriptions of extremes of the types and that most molecules exhibit a mixture of two or three types.

VAN DER WAALS FORCES

The attractive forces between neutral, nonpolar molecules such as hydrogen molecules, or chlorine molecules, or the rare-gas monatomic molecules are given the name of the scientist who first properly identified them. They are the least in magnitude of all the intermolecular forces but extremely important, because they are manifested at all times in all forms of matter. Consider a sample of helium gas. Like all the rare gases, helium is monatomic, which is to say that the gas molecule consists of one atom. Consider two neutral helium atoms. Each consists of a nucleus of two protons and two neutrons, and each has two extranuclear electrons. These electrons are chiefly under the influence of their own nucleus (if this were not so, the concept of the atom would be meaningless), but the electrons are not confined to rigorously defined paths. Thus as the atoms approach each other, the electrons of one atom will—fleetingly—come under the influence of the other nucleus. This situation is repeated throughout the body of the gas continuously. The sum total of these fleeting influences is enough to cause helium to liquefy only when the temperature is below $-268.9°C$. In substances such as chlorine, the larger molecules, with more electrons and higher nuclear charge per molecule, produce van der Waals forces which are sufficient to bring about liquefaction at higher temperatures, at $-34.6°C$ for chlorine, and at $58.8°C$ for bromine, so that bromine is a liquid at room temperature.

POLARITY

Intermolecular attractions are greater between polar molecules than between nonpolar molecules. Van der Waals forces are still operative, but polar attractions may be sufficiently great to overbalance them. Thus we would expect van der Waals forces to be greater between chlorine molecules (34 electrons each) than between CH_3Cl molecules (24 electrons each)—and they are—but the polar attraction of CH_3Cl molecules for each other is sufficient to keep this liquid from boiling until $-24°C$, while liquid chlorine boils at $-34.6°C$.

INTERMOLECULAR ATTRACTIONS

THE HYDROGEN BOND

This strongest of the three intermolecular forces is the least frequently encountered, but where it exists, it is far more important in determining behavior than either of the other two. It arises because of the extremely small size of the hydrogen atom, but it exists only when the hydrogen atom is bonded to one of the small highly electronegative atoms such as oxygen, nitrogen, or fluorine. A further requirement is that the molecule have at least one unshared pair of electrons. The hydrogen bond is formed between the hydrogen atom and the electronegative atom of another molecule in the region of the unshared pair of electrons. Each HF molecule has three unshared pairs but only one hydrogen atom. Each NH_3 molecule has three hydrogen atoms but only one unshared electron pair. The water molecule displays optimum conditions for maximum hydrogen bonding: two unshared pairs and two hydrogen atoms per molecule. Although all three substances exhibit hydrogen bonding (and thus, for instance, have much higher boiling points than might be predicted), the extent of hydrogen bonding in water is so great that its properties appear very abnormal. Placing the hydrogen compounds of the nonmetals in the same periodic group as oxygen in order, starting with the largest molecule, and considering their boiling points, we have the order shown in Table 7-5. These boiling points vary as we would expect: as the number of electrons per molecule decreases, van der Waals forces decrease and the boiling point gets lower and lower. Pursuing this line of reasoning then, we should predict that water, H_2O, would boil at somewhere around −80°C. Instead, as we know, water remains in liquid form until a temperature of 100°C is reached! The extra kinetic energy is required to overcome the hydrogen bonding. This same kind of deviation in boiling point is found with the other hydrogen-bonded liquids mentioned, although not to the same degree. Table 7-6 shows the boiling points for the binary hydrogen compounds of the elements of groups V and VII and also, for comparison, the binary hydrogen compounds of the elements of group IV (CH_4 has no hydrogen bonding).

TABLE 7-5 Boiling Points of Binary Hydrogen Compounds of Some Group VI Elements

Compound	Boiling Point, °C
H_2Te	−1.8
H_2Se	−41.3
H_2S	−59.6

THE PHYSICAL STATES OF MATTER: FLUIDS

TABLE 7-6 Boiling Points of Binary Hydrogen Compounds, °C

Group IV		Group V		Group VII	
SnH_4	−48	SbH_3	−17	HI	−35.5
GeH_4	−90	AsH_3	−55	HBr	−67
SiH_4	−112	PH_3	−85	HCl	−85
CH_4	−161	NH_3	−33.4	HF	19.4

The boiling point of liquids was selected for this detailed discussion of intermolecular forces, but a similar explanation can be developed for other variations in liquid properties.

Thus far, then, the kinetic molecular theory has been successful in explaining the nature of the liquid and the gas states and their properties as well as variation in properties among individual liquids and gases. We shall find in the next chapter, however, that the extension of the theory to the solid state accounts for the properties of only a few solid substances. The solid state of matter involves some completely new relationships.

Questions

1 For which of the three states of matter is the property of odor most characteristic? For which state is color the most characteristic?

2 Name two substances, one a solid and one a liquid, of which the solid is heavier than the liquid. Name another pair of which the liquid is the heavier of the two.

3 It is possible for an *object* to have a density different from that of the material of which it is composed. Explain.

4 Which is the densest gas in Table 7-1? Is this substance also the heaviest solid in the table? Which is the lightest gas? Is this substance also the lightest solid?

5 How many times denser is Dry Ice (solid CO_2) than its vapor form? (Refer to Table 7-1.)

6 List some practical uses made of the fact that gases are compressible.

7 List some uses of the fact that liquids are *not* compressible.

8 Explain in your own words why the liquid in all four tubes of Fig. 7-4 stands at the same level.

9 Indicate the effect upon the volume of a gas if the following changes are made. If it is not possible to make a decision on the basis of the information given, indicate that this is the case:

   a Pressure is increased while temperature and quantity remain constant.

   b Temperature is decreased while pressure and quantity remain constant.

QUESTIONS

   c Quantity is increased while pressure and temperature remain constant.
   d Pressure and temperature are both increased while quantity remains constant.
   e Pressure and temperature are both decreased while quantity remains constant.
   f Pressure is increased and temperature is decreased while quantity remains constant.
   g Pressure is decreased and temperature is increased while quantity remains constant.
10 Calculate the resulting volume at constant temperature if:
   a The pressure on a 10.0-liter sample of gas is increased from 1.00 to 2.00 atm.
   b The pressure on a 10.0-liter sample of gas is reduced from 1.00 to 0.333 atm.
   c The pressure on a 10.0-liter sample of gas is increased from 760 to 1,140 mm.
11 A liter of gas is subjected to the following changes. In each case assume that all other conditions remain constant and determine the new volume.
   a Pressure is tripled.
   b Kelvin temperature is doubled.
   c Pressure is doubled, and Kelvin temperature is tripled.
   d Pressure is halved, and quantity is doubled.
   e Quantity is tripled, and Kelvin temperature is halved.
   f Quantity is doubled, Kelvin temperature is tripled, and pressure is reduced to one-fourth of the original pressure.
12 If 1.00 liter of a gas at STP is subjected to a pressure of 840 mm while the temperature is held constant, what will be the new volume?
13 Suppose two 1-liter containers as in Fig. 7-6 contain gases but the pressure of gas A is 1.00 atm while the pressure of gas B is 2.00 atm. If the partition is lifted so that the two gases can mix freely, what will be the resulting pressure on the whole container?
14 What will be the partial pressures of gas A and gas B (Question 13) after mixing?
15 What is the normal boiling point of water on the Kelvin scale? What is the freezing point?
16 What is the relation between vapor pressure and boiling point?
17 (Use Table 7-2 in answering.)
   a Doubling the Kelvin temperature of 173°K has what effect upon the volume of 0.63 liter?
   b By what fraction has the Kelvin temperature been multiplied when the Celsius temperature is changed from 0.0 to 100°C? Show that

the volume at constant pressure has been multiplied by the same factor.

18 Prepare a chart listing for comparison and contrast the observable characteristics of the two phenomena of evaporation and boiling.
19 The velocity of the hydrogen molecule at ordinary temperatures is approximately 1 mile/sec, yet when a container of hydrogen is opened, it will require several minutes before the presence of hydrogen can be detected in the far corners of a room. Explain.
20 What is meant by an elastic collision?
21 **a** Calculate the molecular weights of the gases CO_2, CO, SO_2, CH_4, Cl_2, H_2.
   b Arrange these gases in order of increasing rate of diffusion.
22 Calculate the gram molecular weight of
   a Water   **b** Ammonia
   c Carbon tetrachloride   **d** Bromine
23 Give the weight of a mole of:
   a Methyl alcohol, CH_3OH **b** Nitrogen gas, N_2
   c Sulfur, S_8 **d** Oxygen, O_2
24 Given a sample of CO_2 weighing 88.0 g:
   a How many moles are in the sample?
   b At STP what volume would the sample occupy?
   c How many molecules does it contain?
25 How many molecules are there in:
   a 64 g of SO_2?   **b** 16 g of O_2?   **c** 36 g of H_2O?
26 At STP 11.2 liters of CO:
   a Contain how many moles?
   b Contain how many molecules?
   c Weigh how much?
27 Given 1.00 mole of hydrogen gas:
   a What is the weight of the sample?
   b How many molecules does it contain?
   c What is the weight in grams of one molecule of H_2?
28 Given a sample consisting of 32.0 g of methane, CH_4:
   a How many moles is this?
   b What volume at STP would this sample occupy?
   c If 16.0 g more of CH_4 is forced into the container in part **b** while the temperature is held constant, what will be the effect upon the pressure? What will be the new pressure?
29 Give examples of equilibrium situations from your own experience.
30 List some practical uses of the thermal (heat) effects of evaporation.
31 Use Table 7-4 to determine what percent of water molecules have high enough kinetic energies to have escaped the liquid state at 90°C.
32 A container of water is set under a large bell jar as in Fig. 7-8. If the

total volume of all gases under the bell jar is exactly 1.00 liter and the total pressure is 840 mm while the temperature is 30°C, what volume of water vapor is present? *Hint:* What percent of the whole gas volume is water vapor?

33 **a** Consider the four substances fluorine, iodine, bromine, and chlorine. What type(s) of intermolecular forces are present in each of these substances?

 b Arrange the four substances in probable order of boiling points from lowest to highest.

34 Describe (identify) the intermolecular forces in the substances Cl_2, HBr, NH_3, CO_2, CH_3Cl.

35 In this text gases and liquids are discussed in one chapter and solids are given separate treatment. Other texts frequently discuss liquids and solids together. What similarities and differences between the three states might be used to justify this alternate arrangement?

36 How would you explain the following?

 a Gases do not follow Boyle's and Charles' laws exactly.
 b Liquids diffuse much more slowly than gases do.
 c Boiling points vary with geographic location.
 d Evaporation causes cooling.
 e The temperature of a boiling liquid does not increase as it absorbs more heat.
 f Energy must be absorbed by a liquid in order for it to evaporate.
 g The liquid in an automobile radiator may begin to boil when the car is climbing a mountain, but it does not boil at the same temperature at a lower elevation.
 h Upon sufficient cooling, gases liquefy.
 i Water can be made to boil at temperatures considerably below 100°C.
 j Insects which are several times denser than water can walk on the surface of a lake.
 k Pressure increases in automobile tires on hot days.
 l Deviations from the gas laws become more pronounced at lower temperatures and high pressures.
 m Drops of water are spherical in shape.
 n There are no gases at 0°K.

CHAPTER 8 THE PHYSICAL STATES OF MATTER: SOLIDS

All true solids have a regular geometric structure. As seen in this photograph of rock salt, the crystalline nature of some salts is apparent to the naked eye. Other solids reveal their crystalline structures only when magnified. (Morton Salt Company.)

CHAPTER 8 THE PHYSICAL STATES OF MATTER: SOLIDS

A description of the properties of gases was rendered quite simple by the fact that in many situations and under ordinary conditions all gases behave identically. Liquid properties, while showing much more variation, are still identifiable as properties of a state. We shall have to be much more selective in identifying properties of the solid state. Some solids, such as sulfur, talc, and chalk, are powdery and soft; others, such as silver and gold, are relatively soft but consist of a single lump which can be shaped by mechanical means. NaCl and $KClO_3$ are obviously made up of small, hard crystals. The diamond is also crystalline but very different from NaCl, common table salt. Wax, glass, and the plastics exhibit properties different from those of any of the other substances mentioned. (It is important to remember that since only pure substances are being considered, many solids which are not homogeneous will not be discussed.)

What properties do all solids exhibit? We have already suggested (Chap. 7) the two most important: (1) the density of the solid state of a substance is generally greater than that of the liquid or the vapor state of the same substance, and (2) solids maintain their volume and shape independent of the container. These two properties, while general, are restrictive enough so that we shall find that they do indeed characterize solids. Properties of hardness, ductility, luster, and even melting points show too much variation among the solids to serve as properties of the state. Microscopic scrutiny of the individual particles of solids reveals that all solids have a crystalline structure. There is a geometric regularity repeated over and over in the structure of the solid which is not only observable but characteristic.

We are seeking, then, an explanation for the relative density of solids and for the geometric regularity in their submicroscopic structure. An explanation for this must also allow for explanation of the variations in other solid properties. It may seem at first that solids present more differences than similarities! It will be helpful to find some way to classify these differences. The following five groups of substances were suggested in the opening paragraph.

Classes of Solids

I *Soft or powdery low-melting solids.* CO_2, S_8, I_2, talc, graphite, naphthalene (moth balls), and at low temperatures N_2, O_2, HCl, He, Ar, SO_2, Cl_2, Br_2, NH_3, H_2O, CH_4, C_2H_6.

II *Crystalline, brittle, and hard high-melting solids.* NaCl, $KClO_3$, $CaCO_3$, $CuSO_4$, MgO, $Pb(NO_3)_2$, Fe_2O_3.

III *Malleable or ductile, lustrous, low-melting solids.* Ag, Pb, Au, Na, Mg, Pt, Al, Ca, Cu, Fe.

IV *Diamondlike, extremely hard, crystalline, and high-melting solids.* Diamond, silicon, silicon carbide.

V *Rubbery or plastic substances.* Wax, glass, plastics, tallow, paraffin.

The examples are not exhaustive but perhaps are extensive enough for the student to answer the type of question posed once before. Remembering that we seek an explanation for properties in the submicroscopic structure of matter, *what submicroscopic characteristic do the members of each class of solids have in common?* The answers to this question provide us with the first clues to the explanation we seek.

Class I is composed of nonmetallic elements and compounds. In each case the fundamental unit of the substance is a molecule. Class II consists exclusively of ionic compounds; thus the fundamental units are ions. Class III contains metallic elements only, and we shall see the fundamental units are metal ions and electrons. For classes IV and V we do not yet have enough information to deduce the nature of the fundamental unit. We can make a beginning with the first three and then perhaps be better able to understand the commonly accepted theories for the structure of members of classes IV and V.

CLASS I: MOLECULAR SOLIDS

The explanation for the properties of this class of substances is found in a logical extension of the kinetic molecular theory. As the temperature of a liquid is lowered, the average kinetic energy of the molecules is reduced to the point at which they are no longer free to move over and around each other but, because of intermolecular forces, are localized with respect to each other. At this point we say the liquid has solidified or frozen. Individual molecules still retain an appreciable kinetic energy and do, in fact, still manifest molecular motion, but it is motion that is best described as vibration, since each molecule is restricted by intermolecular forces to a definite, bounded position in space relative to the others. The reverse process, melting, takes place when a solid is heated until the average kinetic energy of the molecules is sufficient for them to escape the restrictive effects of the intermolecular forces and they begin to flow. Important factors in determining the melting point of a molecular solid are the type and degree of intermolecular forces operative. The strength of van der Waals forces depends upon the number of electrons in the molecule. The strength of polar attractions depends upon the geometry of the molecule, and hydrogen bonding, when present, usually overbalances all else.

Molecular solids exert an appreciable vapor pressure. The odor of moth crystals is evidence for this. Some molecular solids, such as familiar Dry Ice (solid CO_2) have such a high vapor pressure that the entire solid vaporizes when heated instead of melting. This process of passing directly from the solid to the vapor state is known as *sublimation*. It is possible to melt solid CO_2 or any other solid which sublimes under ordinary conditions by increasing the pressure as it is heated, thus preventing vaporization and allowing the solid to melt.

THE STRUCTURE OF ICE

Water, H_2O, is a molecular compound which, as we have noted above, is a solid at low temperatures. When we discussed it as a liquid (Chap. 7), we discovered that its unusual properties were ascribed to the high degree of hydrogen bonding. The fact that the solid form of H_2O is less dense than the liquid form (Table 7-1) is another of the unusual properties of this compound. Table 8-1 shows that the melting points of hydrogen-bonded solids are abnormal, as we found the boiling points to be (Table 7-6). Another abnormal characteristic is shown in Table 8-2, which gives densities of H_2O at various temperatures. From these figures it can be seen not only that ice is less dense than liquid water but also that water at 4.0°C is H_2O in its densest form.

Can we explain these apparent exceptions to normal behavior in terms of the submicroscopic structure we have proposed for molecular solids? Yes, for once again, the extent of hydrogen bonding is the reason.

As liquid H_2O is cooled from its boiling point, the lowered kinetic energy allows the molecules to move closer together—or, stated another way, gives intermolecular forces relatively more importance—and the liquid shrinks slightly and becomes more dense. With most liquids this process continues to the freezing point. But with H_2O (and other extensively hydrogen-bonded substances) another process begins to assume greater relative importance as the temperature drops, namely, the formation of hydrogen bonds. This process has an effect directly counter to

TABLE 8-1 Melting Points, °C, of Binary Hydrogen Compounds of Group V, VI, and VII Elements

Group V		Group VI		Group VII	
SbH_3	−88	H_2Te	−48	HI	−50.8
AsH_3	−113.5	H_2Se	−64	HBr	−86
PH_3	−132.5	H_2S	−82.9	HCl	−111
NH_3	−77.7	H_2O	0.0	HF	−83

TABLE 8-2 Densities of H_2O

Temperature, °C	Density, g/ml
50	0.98807
40	0.99224
30	0.99567
20	0.99823
10	0.99973
4.0	1.0000
0.0	0.99987
Ice at 0.0	0.917

that of the lowered kinetic energy, because as the bonds form, molecules are held at the distance of the bond and prevented from moving closer together. As the temperature decreases, first small aggregates of molecules and then larger networks of molecules are built up with hydrogen bonds (Fig. 8-1). Until the temperature of 4.0°C (3.98°C, exactly) is reached, the effect of reduced kinetic energy predominates in increasing density. Below 4.0°C, increasing hydrogen bonding with the formation of larger aggregates of molecules predominates and density decreases. At the freezing point there is a sudden increase in hydrogen bonding with attendant reduction in density (Fig. 8-2). A very practical result of all this is that our lakes and streams freeze from the top down instead of the bottom up!

The heat of vaporization of a liquid has been discussed (page 137) and defined as the amount of heat necessary to vaporize one gram of a substance at its boiling temperature. A similar property is the *heat of fusion*, or the amount of heat which must be provided in order to melt one gram of a solid at its melting temperature. Water has the highest heat of vaporization and highest heat of fusion of all common substances. The melting and vaporizing processes require these large amounts of energy to break the hydrogen bonds in addition to overcoming the usual van der Waals forces. The large amounts of heat absorbed by water in vaporizing or by ice in melting are released to the surroundings when steam condenses (which explains the severity of steam burns) or when water freezes.

To complete the description of the unusual properties of water, we refer

Figure 8-1 Structure of ordinary water.

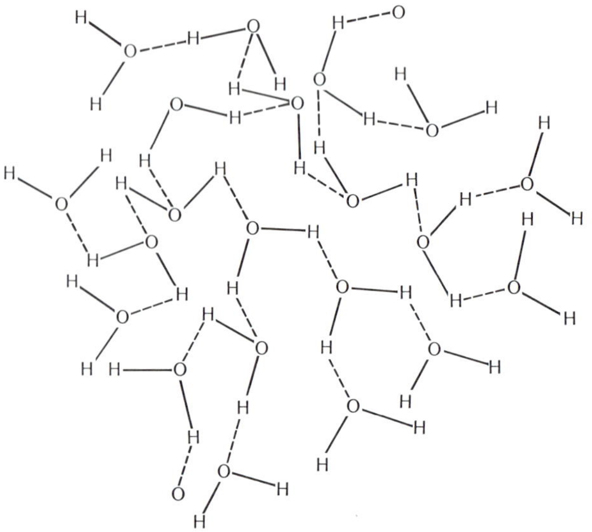

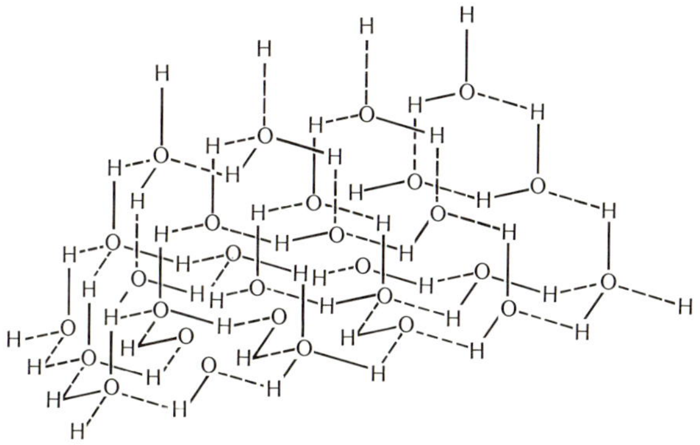

Figure 8-2 Structure of ice.

again to specific heat (Chap. 1). Specific heat is the amount of heat necessary to raise the temperature of one gram of a substance one degree Celsius, and the value for H_2O is abnormally high, again, due to the energy required for breaking hydrogen bonds. The large amount of heat necessary to warm a body of water is released to the surroundings when it cools. This explains the moderating effect large lakes have upon the climate of the countryside.

POLYWATER

Water is the most abundant compound in the world. It has been analyzed in the finest laboratories in the world. Its properties have been determined with the highest degree of accuracy. We have been so confident of its properties that we have used it as the standard for basic units, such as the calorie, the Celsius degree, and the gram.[1] It seems almost inconceivable that any form of water should escape detection until the 1960s. Yet that may have happened. In 1962 a Russian scientist announced a new form of water which has subsequently been called *polywater*.

This substance has properties remarkably different from those of ordinary water. It does not evaporate. At about $-45°C$ it forms a glassy solid, quite unlike ice, without expanding as ordinary water does in solidifying. In the liquid state it is roughly 15 times more viscous than ordinary water. It has no maximum density.

These differences in properties naturally raise the question of how polywater and normal water differ in structure. All evidence seems to indicate that the difference lies in the hydrogen bonding. In normal water

---

[1] The gram was defined in 1798 as $\frac{1}{1,000}$ of the mass of a cubic decimeter of pure water at its maximum density. The present definition of a gram is $\frac{1}{1,000}$ of the mass of a specific cylinder of platinum-iridium alloy kept in France by the International Bureau of Weights and Measures.

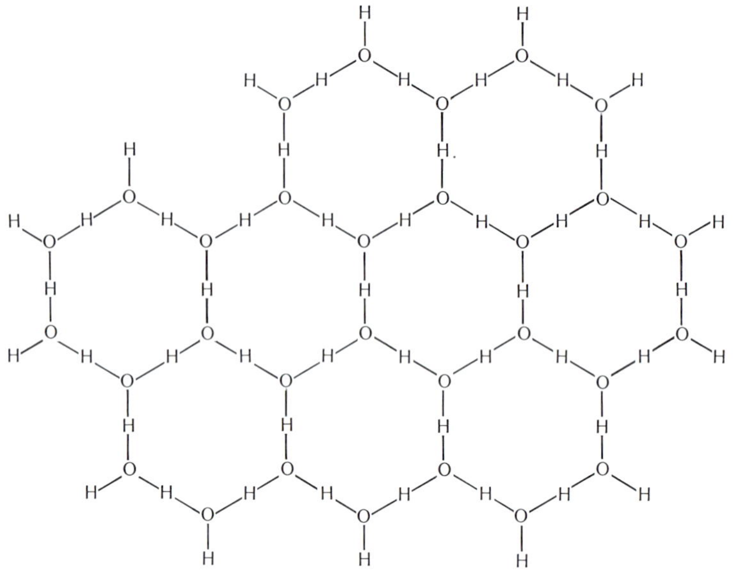

152 THE PHYSICAL STATES OF MATTER: SOLIDS

Figure 8-3 Proposed structure of polywater.

there is considerable difference in bond lengths and strengths (see Fig. 8-1). However, in polywater the O—H—O bonds are of equal length and strength (see Fig. 8-3). The bonds that tie the H$_2$O units together are much stronger in polywater than in ordinary water. In fact, the energy required to break the bonds in polywater is about 20 times that required to break the bonds in normal water.

Some reputable scientists have raised the possibility that, rather than being a new, well-defined form of water, the samples which have been obtained are merely ordinary water containing impurities which have been introduced during its preparation—the impurities being responsible for the altered properties. Only minute quantities of the alleged polywater have been prepared thus far. At present (late 1970) there is a great deal of activity and discussion centered about attempts in both the United States and Russia to prepare polywater in quantity. The debate over its existence may be resolved by the time this book is published.

CLASS II: IONIC SOLIDS

The structure of ionic compounds has been briefly referred to in Chap. 4. The transfer of electrons from one atom to another results in the formation of charged particles called ions. Ions are held together by electrostatic forces in aggregates called *ionic crystals*. The electrostatic attractions of a positive particle for a negative one and of a negative particle for a positive one are extended equally in all directions. Thus each positive ion is surrounded by negative ions, and each negative ion is surrounded by positive ions in a regular, repeated geometric arrangement.

The electrostatic, or ionic, bond is a strong chemical bond, many times stronger than the total of the intermolecular forces which maintain the structure of a molecular solid. As a result, ionic crystals are hard and brittle while molecular crystals are soft and easily crushed. Melting points for ionic crystals are among the highest for pure substances.

CLASS III: METALLIC SOLIDS

An explanation of the characteristics of metallic solids may begin with the contrast between them and the rare gases, which solidify *only* at very low temperatures. This is because the weak van der Waals forces between the monatomic molecules are the only intermolecular forces opposing the kinetic energy of the molecules. Van der Waals forces have been described as the total of the momentary influences of one nucleus upon the electrons of another atom or molecule. It is easy to see that these forces should be at a minimum between atoms consisting of filled energy levels, as in the rare gases. Metal atoms, however, present a very different situation. Each metal atom has only one, two, or three electrons in its outermost energy level. These electrons, called valence electrons (see Chap. 4), are attracted by the nucleus with much less force than are the other electrons of the atom. Thus, in an aggregate of metal atoms the influences of other nuclei upon the valence electrons are relatively more important. In fact, to explain the properties of metals, the theory of the metallic bond assumes that these valence electrons become completely delocalized. It is postulated that instead of each set of valence electrons remaining in the vicinity of its own nucleus most of the time, all valence electrons are equally attracted to all nuclei, thus binding the aggregate together. The metallic crystal is pictured as consisting of positive metal ions held together by these attractions for a "sea" of electrons, the latter composed of the valence electrons.

CLASS IV: NETWORK SOLIDS

The diamond is the extreme representative of this class. Consisting of pure carbon, the diamond is the hardest substance known. It exhibits a crystalline form and has an extremely high melting point. All these facts would lead one to suspect the presence of very strong chemical bonds between the atoms of carbon. The only chemical bond which meets all these requirements is the covalent bond (see page 96 and Fig. 6-5). Since there is no finite unit of the diamond other than the carbon atom, the entire crystal is considered to be the molecule.

CLASS V: RUBBERY OR PLASTIC SUBSTANCES

No matter how intense the scrutiny they are subjected to, none of the substances in this last class show the characteristic geometrically regular structure present in all other solid substances. In fact, these are not solids at all. Some refer to them as *pseudo solids*. A better designation is *supercooled liquids*. They are liquids which have been cooled to a point below their normal freezing temperature. At this temperature the flow or movement of molecules takes place at a very reduced rate, but the molecules have not yet taken up the localized positions which will give the substance a geometric structure. This will take place if these supercooled liquids are held below their melting points for a sufficient length of time. Glass becomes brittle with age, an indication of the fact that it has crystallized. Some texts refer to substances of this type as *amorphous solids* (amorphous meaning without form or shape), but if we are to take the characteristic of having a definite geometric shape to be one of the principle, characterizing properties of true solids, we shall have to avoid this self-contradictory term.

The student may remember hearing the term amorphous applied to the powdery form of carbon known as lampblack, boneblack, or charcoal. Amorphous carbon consists of particles so small and so finely divided that the property of crystalline structure is negligible in its effect upon most of the physical properties of the solid. However, this so-called amorphous substance is a true solid, for it has a definite, characteristic melting point, whereas the rubbery, plastic substances to which the same term is sometimes applied do not. With the application of heat, these latter substances simply become progressively less viscous until they are recognizable as liquids.

We have, then, only *four classes* of solid substances, and each involves a regular geometric shape as a characteristic. We next examine the nature of this geometric shape more closely.

Crystal Lattices and Properties of Solids

The crystals of solids may be composed of atoms, ions, or molecules. The forces between these particles range from very weak van der Waals forces to the covalent bond. All these variations have their effects upon the nature of the solid crystal. However, it has been found that although there are many arrangements possible, most natural crystals consist of one or another of a relatively few geometric patterns.

In the following figures and diagrams the atom, ion, or molecule of

CRYSTAL LATTICES AND PROPERTIES OF SOLIDS

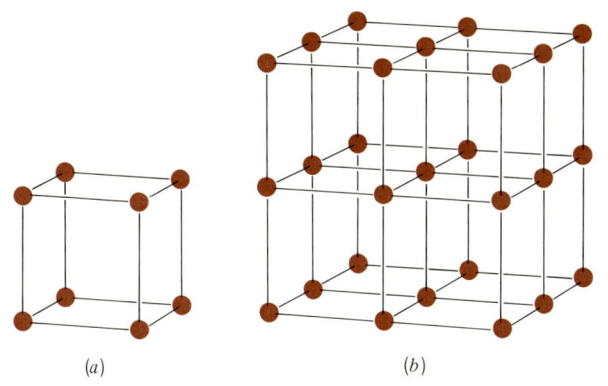

Figure 8-4 Simple cubic crystal: (*a*) unit cell; (*b*) lattice.

the solid is designated by a circle, no attempt being made to differentiate among different kinds of particles. The circle is used to designate the relative position of the particle and sometimes its relative size. The student should keep in mind while studying these diagrams that atoms, ions, and molecules have neither a definite size nor a specifically bounded shape. We are here disregarding the properties of the particle in order to concentrate on the pattern of the arrangement.

Crystal patterns are identified by the *structural unit,* that is, by the geometric unit of structure which when repeated parallel and adjacent to itself generates the entire crystal. The *simple cubic* unit is one such structural unit. Note how the repetition of the cube (Fig. 8-4*a*) in all directions parallel to itself generates the larger crystal pattern in Fig. 8-4*b*. This larger crystal pattern is called a *crystal lattice* and refers only to the geometric arrangement of atoms, ions, or molecules in the solid.

Three other common structural units are shown in Fig. 8-5.

In the simple cubic, face-centered cubic, and hexagonal crystals, each atom, ion, or molecule has 12 nearest neighbors, or particles immediately

Figure 8-5 Other unit cells: (*a*) body-centered cubic; (*b*) face-centered cubic; (*c*) hexagonal.

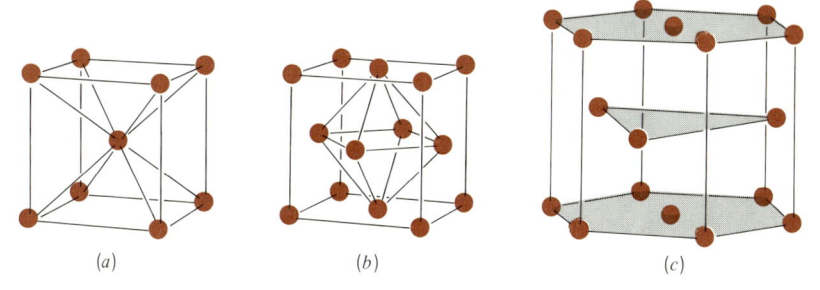

adjacent. In the body-centered cubic structure each particle has only 8 nearest neighbors, and it is thus somewhat less dense than the others. The simple cubic, face-centered cubic, and hexagonal structures are all designated as *closest-packed* structures. It should be apparent how crystal structure affects the relative density of the solid.

Let us next choose a solid from each of the four classes, all four crystallizing in the same lattice, and examine the relationship between response to mechanical stress and the nature of the particle at the lattice points. All four examples chosen below crystallize in the simple cubic closest-packed lattice.

Hydrogen, H_2, is a molecular solid. All lattice points are occupied by hydrogen molecules. Forces between molecules are exclusively van der Waals forces, nondirectional in nature and relatively weak. Solid hydrogen is soft and easily crushed because with a little pressure a hydrogen molecule can slip from one lattice position to another which it occupies as easily as it occupied the original one. Thus whole units or layers may slide, one over another, with ease.

Sodium chloride, NaCl, is an ionic solid. Lattice points are occupied alternately by positive and negative ions (see Fig. 8-6). One layer cannot slide past another in response to pressure, or like ions would be brought next to each other. The strong electrostatic attractions between unlike ions and equally strong electrostatic repulsions between like ions cause the NaCl crystal to be hard and to resist deformation up to the point of fracture. In an ionic crystal lattice, such as in the NaCl crystal, there is one direction in which deformation can take place with somewhat less of an energy requirement. Figure 8-7 shows that layers taken in a diagonal direction through the crystal consist of one kind of ion only. These layers, then, can be moved relative to each other without encountering strong electrostatic repulsions. This fact is responsible for the *cleavage* of crystals, although the energy requirements are frequently high.

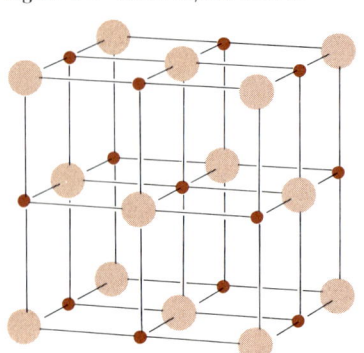

Figure 8-6 NaCl crystal lattice.

CRYSTAL LATTICES AND PROPERTIES OF SOLIDS

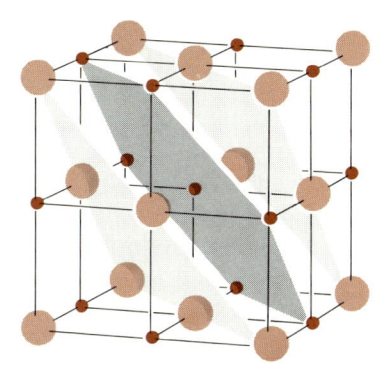

Figure 8-7 NaCl crystal lattice showing diagonal layers containing similar ions.

Copper, Cu, is a metal. The metallic bond, described as an extreme manifestation of van der Waals forces is many times stronger than the ordinary van der Waals forces. Evidence for this statement is found in the strength of metals. But many metals can be deformed quite easily. They can be bent, drawn, and hammered thin without breaking. If we consider the nature of particles at the lattice points the reason for this becomes apparent. As in the molecular crystals, all lattice points are occupied by the same kind of particle, so that slipping one layer over another is not impeded by electrostatic forces.

Diamond is a network solid with carbon atoms held in the lattice points by strong, directional covalent bonds. Any distortion at all requires the breaking of these bonds and results in fracture but only with the expenditure of relatively large amounts of energy.

ELECTRIC CONDUCTANCE

The different ways solids respond to the application of an electric current can be explained by the theories of submicroscopic structure already outlined. Metals exhibit the property of conducting an electric current and are called *conductors*. Other types of solids are called *insulators*. A current of electricity consists of a movement of electrons. The sea of electrons present in the metallic crystal provides the highly mobile electrons necessary for conductance. In none of the other types of solids are they available.

The graphitic form of carbon might appear to be an exception to this statement. Graphite is a nonmetal and is also an electric conductor. Reference to Fig. 6-7 will help to explain this, however. The pi bonds within the layers of graphite are delocalized, that is, not restricted to one position, which is why a resonance formula for graphite is required. The mobile electrons of these pi bonds are responsible for the conductance of graphite.

The delocalized electrons of the metallic crystal also furnish an explanation for the luster characteristic of metallic surfaces. When light strikes the surface of a metal, it causes the loosely held valence electrons to vibrate, sending back light of almost the same frequency.

The behavior of ionic crystals when melted is worth noting in this context. When sufficient heat has been supplied to break down the crystal lattice, the ions are able to flow. Being charged particles, when they are subjected to an electric potential, the positive ions move toward the negative terminal and the negative ions toward the positive terminal and carry the current. This property of fused, or melted, ionic crystals will be examined in greater detail in Chap. 11.

Summary   With this chapter we conclude the discussion of the structure of matter. We have sought an explanation for the properties of substances in theories of the submicroscopic structure of matter. This has taken us from a consideration of the fundamental particles (electron, proton, and neutron), through descriptions of atoms, ions, and molecules to an examination of the three physical states of gas, liquid, and solid. We are ready, now, to turn to a consideration of the transformations of matter and the means by which such transformations are brought about.

Questions

1 The first sentence of this chapter says that "in many situations and under ordinary conditions all gases behave identically." What kind of behavior is referred to? Under what conditions do gases *not* behave identically?

2 Name some solids which are not pure substances.

3 In which of the four classes of solids would you place each of the following?

NO_2 SiO_2 P_4O_{10} Rn   Hg
CCl_4 Rb F_2 Zn SO_3
$BaBr_2$ Ba   Cr $Mn(OH)_2$ Kr
$(NH_4)_2SO_4$ $SrSO_3$ Co $Zn_3(PO_4)_2$
K   HBr   Ne   Xe

4 Explain in your own words the terms "localized," as in "the molecules become localized," and "delocalized," as in "the electrons are completely delocalized."

5 Carbon dioxide is more frequently transported in the liquid state than in the solid state. What advantages and disadvantages would there be to each of the three physical states in considering the economy of transportation of carbon dioxide?

QUESTIONS

6 Make a list of the unusual properties of water. "Unusual" here means unusual with respect to what?

7 Give examples of other practical results (in addition to the way lakes freeze) of the fact that ice is less dense than water.

8 Explain in your own words how a lake affects the climate of the surrounding area.

9 Of what practical importance are the abnormal melting and boiling points of water?

10 Why is the term "amorphous solid" not appropriate as a designation for wax, glass, and other plastic or rubbery substances?

11 Suggest reasons based upon submicroscopic structure for the following facts:
  a H_2 melts at $-259°C$, but O_2 melts at $-218°C$.
  b HCl melts at $-111°C$, but HI melts at $-50.8°C$.
  c HCl melts at $-111°C$, but HF melts at $-83°C$.
  d Cl_2 melts at $-101°C$, but HBr melts at $-86°C$.

12 a Copper and silver metals both crystallize in a cubic closest-packed lattice. How can you account for the greater density of silver?
  b The density of aluminum is 2.70 g/cm^3, and that of rubidium is 1.53 g/cm^3. What does this imply about their respective crystalline arrangements?
  c What *two* factors must be considered in predicting the probable density of a metal from its submicroscopic structure?

13 Both molecular solids and metallic solids consist of only one type of particle at the lattice points. How is the much greater strength of the metallic bond explained?

14 The diamond also has only one type of particle at all lattice points. Why is no deformation of the diamond possible by one atom of carbon's slipping past another?

15 Compare the solid, liquid, and gaseous states of a molecular substance with respect to:
  a Distance between molecules
  b Velocity of the molecules
  c Energy of the molecules
  d Freedom of movement of the molecules

16 What type or types of intermolecular forces are operative in the following types of solids?
  a Molecular   **b** Ionic
  c Metallic   **d** Network

17 How does the degree of hydrogen bonding in water vary with the temperature?

18 Account for the fact that SiO_2 is a high-melting solid but CO_2 is volatile.

19 How would you explain the following?
   a Ice disappears during the winter even when the temperature remains constantly below the freezing point.
   b Metallic crystals are malleable, but any attempt to deform an ionic crystal shatters it.
   c The temperature of the water at the bottom of a deep lake in winter is likely to be 4°C.

CHAPTER 9 ION EXCHANGE

In an attempt to make white fabrics "whiter than white" and colored fabrics "brighter than bright," manufacturers have developed extremely effective detergents. Unfortunately, the nonbiodegradable synthetic detergents created foaming rivers, such as the one shown in this picture, which are serious environmental problems. This form of water pollution can be eliminated by shifting to soap or biodegradable detergents, that is, detergents which organisms can readily attack and reduce to simple molecules. (Mr. John W. Alley, University of Wisconsin Photography Department, Milwaukee.)

CHAPTER 9 ION EXCHANGE

Having completed our discussion of the composition and structure of matter, let us now turn our attention to its transformation. Changes are constantly taking place in the universe about us; we are not aware of many of the transformations of matter which take place on other planets and distant stars. In a later chapter, however, we shall discuss briefly some of the transformations of matter that take place on the sun.

We are aware of many changes taking place in our own immediate environment. Metal corrodes, water evaporates, dead organic material decays, milk sours, sugar dissolves, etc. We have previously learned to classify these changes as either physical or chemical.

We have already discussed the transformation of matter through changes in physical state. These changes are usually thought of as *physical changes*. In this chapter our concern will be principally with *chemical changes*, that is, changes which produce new substances so that they have different sets of properties and different sets of reactions. As an example of a chemical change we might recall the interaction between metallic sodium and gaseous chlorine to form sodium chloride (see Chap. 4). Sodium chloride, the product of the reaction, has none of the properties of the two elements which reacted to form it.

At this point a word about the limitations of definitions seems appropriate. We have stated that water free from all suspended and dissolved materials is a pure substance and that the changes from ice to liquid water to steam are physical changes. If we think about this statement for a moment, we can see that both parts of it are debatable.

We learned that water does not consist of discrete H_2O particles but H_2O units are held together by hydrogen bonding to form $(H_2O)_n$ molecules, where n can be a small or large number that varies with the temperature. Thus water is a mixture of various kinds of molecules. We shall learn in the next chapter that even pure water contains ions! This being the case, one could argue that pure water is a mixture and not a substance. When one changes the temperature of water, hydrogen bonds are formed and broken. Thus the composition of water changes in terms of the kinds of molecules present; furthermore, certain different properties appear. Therefore, one might argue that changing the temperature of water is a chemical, not a physical, change. The classification of water as a substance can be defended on the basis that the ratio of two hydrogen atoms to one oxygen atom never varies. Also one might argue that changing the temperature of water does not vary the ratio of hydrogen to oxygen, and thus that the changes observed are physical. This closer scrutiny of several definitions reveals that definitions are at best only approximations. Scientists do their best to describe nature, but language and mathematics are often inadequate to the task of making a wholly accurate statement about it.

Why Reactions Occur

Some reactions occur simply when substances are brought together. For example, the reaction between sodium and chlorine takes place with the evolution of considerable energy when chlorine comes in contact with sodium. We have explained *combination* reactions between elements like this as the result of the tendency of atoms to acquire stable electron configurations. The drive of systems for stability seems to be the propeller for all transformations of matter.

Two factors are involved in the drive of systems for stability. One is the amount of energy stored in a system (potential energy). In general, other things being equal, the system with the smallest energy content is the most likely to be stable. Examples of potential energy include rocks at the peak of a mountain, wound-up clocks, compass needles, and water at the top of a dam. All these systems tend to carry out a mechanical process resulting in a loss of energy content. Rocks roll down the mountainside, but there is no tendency for rocks to roll back up the mountain. Similarly, a clock unwinds when measuring time, but there is absolutely no tendency for a clock to wind up its spring again.

In the chemical realm, this tendency to reduce the potential energy of a system is equally evident; chemical energy is, in fact, one form of potential energy. Again, using the reaction between sodium and chlorine as an example, we note that neither of these elements exists in nature in a free state. The reason is that our environment contains materials which react with these substances to liberate chemical energy, but once the reaction has occurred, there is no tendency for the process to be reversed and the potential energy to be restored. Most chemical reactions proceed with the liberation of energy; that is, they are *exothermic*.

However, other reactions are *endothermic*. That is, they proceed by using energy. For example, if energy is supplied to sodium chloride in the right form under the right conditions, the compound can be changed back into the elements. This reaction takes place because of the tendency of a system to increase its *entropy*. This is a tendency of particles to become as chaotic, mixed up, or random as possible. We can probably best understand the concept of entropy by considering some physical processes first. Consider the experimental setup in Fig. 9-1. The apparatus contains samples of two gases in containers of equal size at the same temperature and pressure. When the stopcock separating the two samples of gases is opened, the gases quickly mix to form a uniform mixture (a uniform mixture is referred to in chemistry as a *solution*). That is, a maximum amount of randomness or "mixed-upness" is quickly reached. No potential energy is lost as a result of forming this solution. The driving force in this physical process is the tendency of the system to acquire the highest degree of entropy.

This tendency to form solutions can also be illustrated with similar liquids. In the experimental setup in Fig. 9-2 two liquids are in the same

WHY REACTIONS OCCUR

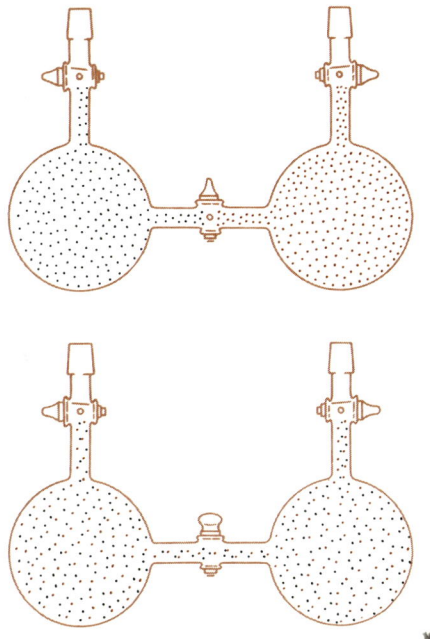

Figure 9-1 The highest degree of entropy is acquired in a system involving gases when a uniform mixture is formed.

container. The lighter is carefully floated on top of the heavier one. In time the two liquids randomly diffuse to form a solution. No potential energy is lost as a result of forming the solution. As in the case of the gases, there is no tendency for the reverse process; that is, the molecules do not return to an ordered pattern but remain in the maximum chaotic state. Again, solution formation is ascribed to the tendency of systems to achieve maximum entropy.

In some instances the two factors, loss in potential energy and gain in entropy, work in the same direction. In others they oppose each other.

Figure 9-2 The highest degree of entropy is acquired in a system involving similar liquids when a uniform mixture is formed.

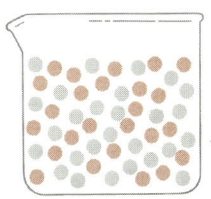

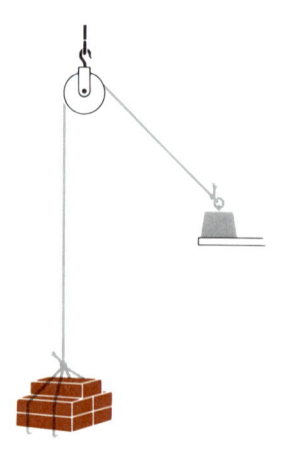

Figure 9-3 An object on the shelf possesses potential energy.

The combination of these two factors is called free energy. All systems (whether a falling object or a sodium-chlorine reaction) tend to minimize free energy. Free energy becomes smaller when potential energy decreases or when entropy increases.

Free energy can be defined as the ability to do work. Let us consider the system of the falling object. If the weight in Fig. 9-3 is pushed from the shelf, it loses potential energy; hence the free energy of the system decreases. Also the bricks on the other side of the pulley are raised (assuming, of course, that the weight is sufficiently heavy) and work is accomplished (see Fig. 9-4). In its new position the weight has less ability to do work. Note, however, that the free energy would have decreased by the same amount if the weight had been allowed to fall freely and whether or not the bricks were lifted. Free energy is the *ability* to do work; it is immaterial whether or not, when the change in the system takes place, work is actually accomplished.

Figure 9-4 The falling weight does work, and the free energy of the system decreases.

Similarly, in the sodium-chlorine system, work may (or may not) be accomplished as the free energy in the system is decreased. If the heat liberated by the reaction is used to boil water, the steam can be used to run a steam turbine. After all the sodium and chlorine have formed sodium chloride, the system has less ability to do work. Remember that the decrease in free energy takes place regardless of whether the heat is used to perform useful work or is allowed to dissipate into the surroundings.

Chemical reactions occur by the transfer of particles between substances. In the reaction between sodium and chlorine, the particle transferred is the electron. Other particles transferred in a chemical reaction are extranuclear protons and ions. We shall consider the transfer or exchange of ions in this chapter and leave other categories for a later chapter.

THE REACTION BETWEEN AgF AND NaI

Let us consider the exchange of ions between crystals of AgF and NaI. That is, we wish to change AgF and NaI into AgI and NaF by an exchange of ions. How do we accomplish this? Our first inclination is to try mixing crystals of the two compounds and hope that, as in the case of sodium and chlorine, contact between the substances will bring about the desired reaction. Unlike the sodium and chlorine mixture, however, the mixture of AgF and NaI is a stable system. Nothing happens when the two kinds of crystals are mixed, but if a third substance, water, is added, the desired chemical reaction takes place. The white crystals of AgF and NaI quickly disappear, and the light yellow crystals of AgI form. The Na^+ and F^- ions remain in solution. Crystals of NaF can be obtained by filtering the mixture and evaporating the filtrate (see Fig. 9-5).

The Solution Process

Let us take a closer look at the reaction between NaI and AgF. The crystals of AgF and NaI quickly disappear because they form a homogeneous mixture, or solution, with water. We say these compounds are soluble in water. However, AgI does not form a homogeneous mixture, or solution, with water. We say that it is insoluble in water. In other words, Ag^+ and I^- cannot exist together in solution. When NaI and AgF go into solution, the Ag^+ and I^- form new crystals of AgI and the Na^+ and F^- ions remain in solution.

Why do NaI and AgF dissolve? Why are Ag^+ and I^- ions incapable of existing together in solution? We can answer these questions, at least in part, by considering what takes place in the solution process. Let us consider the first question. When NaI comes in contact with water, an interaction between the ions of the crystal and water molecules takes

168 ION EXCHANGE

Mixture of AgI crystals and NaF solution

AgI crystals remain on filter paper

NaF solution

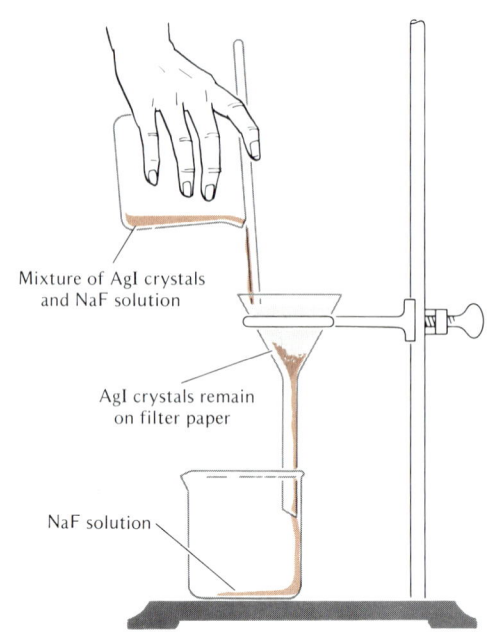

Figure 9-5 Separation of NaF solution from AgI crystals by filtration.

place. The polar water molecules attract and surround the ions of the crystal (see Fig. 9-6). The Na^+ ions are attracted to the negative ends of the water molecules, and the I^- ions are attracted by the positive ends of the water molecules. The name for this process is *hydration*. The attraction of water molecules is sufficiently large to overcome the interionic attraction, and the ions are broken off from the crystal and move out into the body of the liquid surrounded by water molecules.

In terms of energy relationships, crystals of NaI dissolve because free energy is decreased in the solution process. In order for sodium iodide to dissolve, energy must be provided to pull the sodium and iodide ions apart (*crystal energy*). This is an endothermic process. Pulling the water molecules apart, that is, breaking the hydrogen bonds, is also an endothermic process. But the attachment of the water molecules to the ions is an exothermic process: energy is liberated, called the *hydration energy*. The hydration energy is larger than the energy required to overcome the interionic attraction within the crystal combined with the energy required to pull the water molecules apart.

The breakdown in the crystal lattice into the randomness of the solution results in a large increase in entropy. However, this large increase in entropy is partly offset by the decrease in randomness in water mole-

169 THE SOLUTION PROCESS

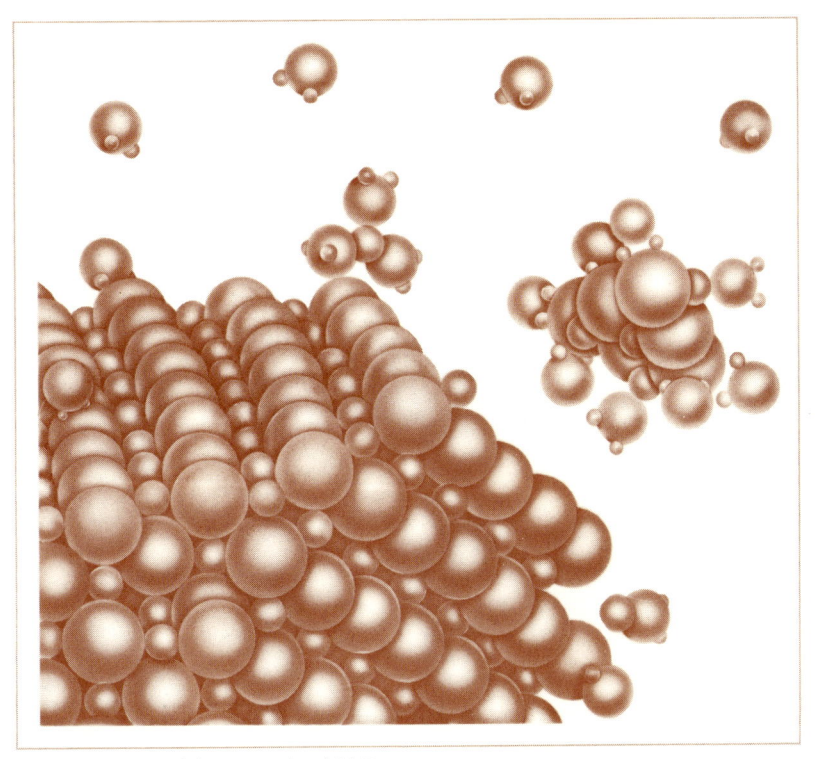

Figure 9-6 Dissolving crystals of NaI.

cules due to hydration. Nevertheless, the net entropy change is a positive one. A schematic representation of particles in solution is shown in Fig. 9-7. Both the potential energy and the entropy factors favor solution: potential energy decreases, and entropy increases.

Let us now consider our second question. Why are Ag^+ and I^- incapable of existing together in solution; in other words, why is AgI insoluble? In this case the interionic attractions are so great that they overcome the attraction of the water molecules for the ions, which tends to keep the ions apart. In terms of energy relationships, entropy would increase if solution took place, but so would the potential energy of the system. Although the entropy change favors solution, the increase in potential energy does not. The entropy change is not sufficient to counteract the potential-energy change; hence free energy would increase upon solution. Therefore AgI is insoluble.

Sometimes the potential energy of the system increases as the crystal dissolves. Such is the case with NH_4Cl. With many solutions the heat produced as a result of a decrease in potential energy results in an appre-

170 ION EXCHANGE

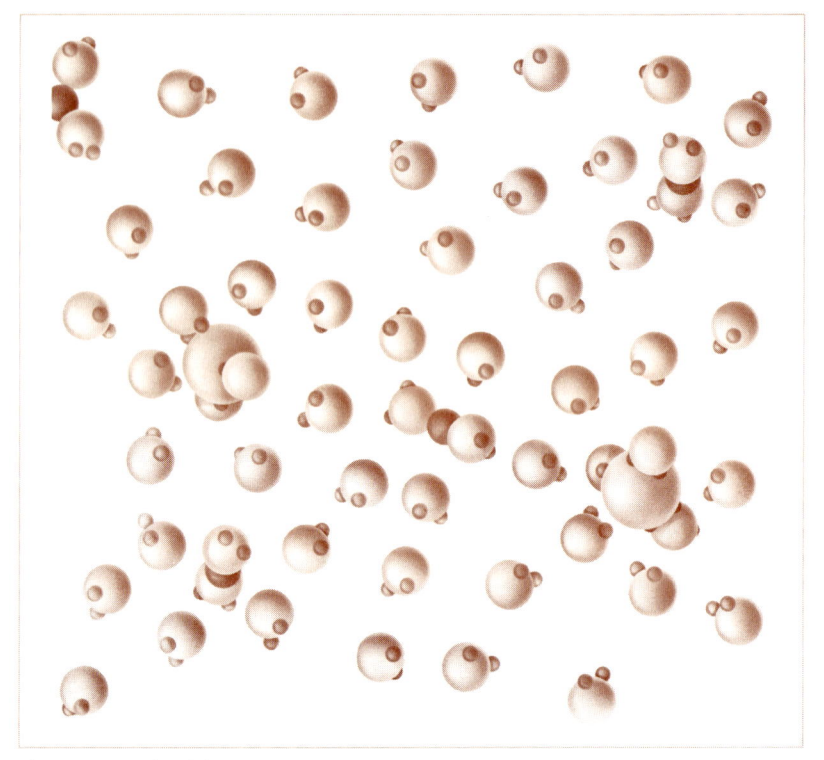

Figure 9-7 Dissolving NaI crystals in water decreases the randomness of the water molecules.

ciable increase in the temperature of the solvent. With NH_4Cl the solvent is cooled because heat is taken from the solvent, increasing potential energy. This compound nevertheless dissolves readily because the increase in entropy is sufficient to overcome the potential-energy change. On the other hand, it can be shown by calculations that calcium fluoride is insoluble because the entropy change upon solution would be negative and the potential-energy decrease would not be enough to counteract the entropy factor. The decrease in entropy which results when calcium fluoride dissolves means that the increase in randomness of the Ca^{++} and F^- ions is more than counteracted by the decrease in randomness of the water molecules. That is, the water molecules become more orderly in CaF_2 solutions than they were in pure water.

SOLUTION TERMINOLOGY

The reader should realize that the terms *soluble* and *insoluble* are relative. Although we have referred to AgI and CaF_2 as insoluble, they *do* dissolve to a slight extent. However, the amount of AgI or CaF_2 that dissolves is very small in comparison with the quantities of AgF and NaI that dissolve in the same amount of water at the same temperature. There is no general agreement on the demarcation line between soluble and insoluble, but a rule of thumb often used is that compounds whose solubility is less than 10 g/liter of solution are considered insoluble. Neither CaF nor AgI dissolves to anywhere near that extent, while the solubilities of AgF and NaI greatly exceed that value.

The components of a solution are referred to as the solute and the solvent. The *solvent* is the dispersing medium, and any other components are the *solutes*. The solvent is present in the largest amount, generally has the same physical state as the solution, and is generally less reactive than the solute. The unique solvent properties of water make it the most widely used medium for chemical reactions. Although a great deal of chemistry is carried out in solvents other than water, water remains the most widely used of all solvents.

In discussing the solubility of NaI and other compounds, we said nothing about the temperature at which the dissolving takes place; we assumed room temperature. The solubility of many compounds is greatly influenced by changes in temperature. In general, solubility increases with increase in the temperature; however, this is not always true.

A solute can be dissolved, a small amount at a time, to give a series of solutions, each containing more solute per amount of solvent than the preceding one. However there is a limit to this process. When a point is reached when more solute cannot be dissolved at a given temperature, we say the solution is *saturated*. Solutions prepared before saturation is reached are referred to as *unsaturated* solutions. When the temperature is raised, the solubility of the solute will likely increase, the solution which was formerly saturated becomes unsaturated, and more solute can be dissolved. If this solution is cooled down to the former temperature *without* any solute's coming out of solution, we have a *supersaturated* solution. Supersaturated solutions are quasi-stable and difficult to prepare. Only a few ionic solids form supersaturated solutions in water, among them are $Na_2S_2O_3$ and $NaC_2H_3O_2$.

When a solution becomes saturated, the solution process does *not* stop. The solute continues to go into solution, but the solute comes out of the solution at the same rate. In the case of an ionic solid, the ions leave the crystal and go into solution at the same rate at which they return to the crystal and are deposited on the surface. That is, an equilibrium is established. Prior to saturation, solution takes place at a faster rate than crystallization, and crystal disappears.

The three types of solutions—unsaturated, saturated, and supersaturated—can be distinguished by adding a small crystal of the solute. If the solution is unsaturated, the crystal dissolves. If the solution is saturated,

172 ION EXCHANGE

nothing happens. If the solution is supersaturated, crystallization takes place until the quantity of solute in the solution is reduced to that of a saturated solution.

Since solutions do vary in the amount of solute, we need some method of describing each solution in terms of the amount of solute present, that is, its concentration. Of the many methods for doing so we shall discuss only one. It is called *molarity* and is the most useful and commonly used method of expressing concentration. Molarity expresses the number of moles of solute per liter of *solution* and is represented by M. Thus, a solution containing 2.00 moles of solute per liter of solution is a 2.00 M solution.

We have defined the mole as that quantity which contains 6.02×10^{23} units. If we are discussing a covalent compound, 1 mole is 6.02×10^{23} molecules. If we are discussing an ion, a mole is 6.02×10^{23} ions. If we are discussing an ionic compound, a mole is 6.02×10^{23} formula units. A formula unit consists of the same ions that appear in the formula in the same number that each ion appears. Thus a formula unit of NaCl consists of one Na^+ ion and one Cl^- ion, and a mole of NaCl consists of 6.02×10^{23} Na^+ ions and 6.02×10^{23} Cl^- ions. A formula unit of $CaCl_2$ consists of one Ca^{++} ion and two Cl^- ions. A mole of $CaCl_2$ contains 6.02×10^{23} Ca^{++} ions and $2 \times 6.02 \times 10^{23}$ chloride ions.

The following examples will help the reader understand molarity and become familiar with it.

Example 9-1 How many moles of solute are there in 5.00×10^2 ml of a 3.00 M solution?

Solution Since the solution is 3.00 M, there are 3.00 moles in each liter. Hence in $\frac{1}{2}$ liter there would be 1.50 moles.

Example 9-2 How many grams of a solute are needed to prepare 2.00 liters of a 0.400 M solution? The solute has a molecular weight of 32.0.

Solution Each liter would require 0.400 mole; 2.00 liters would require 0.800 mole. The number of grams is 0.800 times the number of grams per mole, or $0.800 \times 32.0 = 25.6$ g.

Example 9-3 What is the molarity of a solution prepared by dissolving 242 g of a solute in 2.00×10^2 ml? The molecular weight of the solute is 342.

Solution Since 2.00×10^2 is $\frac{1}{5}$ liter, the number of grams dissolved in 1.00 liter is $5 \times 242 = 1,210$ g. The number of moles in 1,210 g is 1,210/342, or 3.54. Thus there are 3.54 moles in 1.00 liter of solution. This is the molarity of the solution.

173 CHEMICAL EQUATIONS

Chemical Equations

A chemical reaction is often stated in terms of a chemical equation. The equation expresses what is reacting and what is produced in symbols and formulas rather than in words. The first step in writing a chemical equation is to learn the facts about the chemical reactions to be expressed. Since we have just begun our study of the transformation of matter, we do not know many facts about chemical reactions. However, we do know that when AgF and NaI are brought together in solution, AgI precipitates. This reaction is represented by the equation

$$Ag^+ + I^- \longrightarrow AgI$$

In this equation, or any equation, the species reacting (*reactants*) are placed to the left of the arrow and separated by a plus sign. The substances produced (*products*) are placed to the right of the arrow and, if there is more than one, also separated by plus signs.

The reader may wonder about the absence of the Na^+ and F^- ions, in this equation. These species do not appear because they do not take part in a change regarded as chemical, since no distinct new chemical substance is formed. They remain in solution; hence they do not form a new crystal.

The chemical facts regarding the precipitation of common ions from solution can be summarized by a few simple statements:

1. The NO_3^-, ClO_3^-, and $C_2H_3O_2^-$ ions exist in solution with most positive ions.
2. The Na^+, K^+, and NH_4^+ ions exist in solution with most negative ions.
3. The Cl^-, Br^-, and I^- ions exist in solution with most positive ions except Ag^+ and Pb^{++}.
4. The SO_4^{--} ion exists in solution with most positive ions except Ca^{++}, Ba^{++}, and Pb^{++}.
5. The S^{--}, CO_3^{--}, PO_4^{3-}, and OH^- ions almost never exist in solution with any ion of a metal except those of group IA.

With these chemical facts before us, we can write many chemical reactions. Let us consider the reaction between solutions of K_3PO_4 and $Ca(NO_3)_2$. A mixture of these two solutions would bring K^+, Ca^{++}, NO_3^-, and PO_4^{3-} ions in contact. We see from the above statements that Ca^{++} ions and PO_4^{3-} ions will form $Ca_3(PO_4)_2$. Note that the formula requires two PO_4^{3-} ions to three Ca^{++} ions. This is true since all charges in the equation must be neutralized. That is, it requires three Ca^{++} ions, each with a +2 charge, to provide six +'s to neutralize the six −'s on the two PO_4^{3-} ions. The equation for this reaction is

$$3Ca^{++} + 2PO_4^{3-} \longrightarrow Ca_3(PO_4)_2$$

The numbers before the Ca^{++} and PO$_4^{3-}$ ions are called *coefficients*. They are necessary to *balance* the equation. The same number of calcium atoms, phosphorus atoms, and oxygen atoms must appear on each side of the equation. Also the charge must be the same for the two sides of the equation. In this case the charge on both sides is 0. This is accomplished with a 3 in front of the Ca^{++} and a 2 in front of the PO$_4^{3-}$. The reader should note that no coefficients are necessary to balance the equation Ag$^+$ + I$^-$ $\longrightarrow$ AgI.

Chemistry of Soap and Detergents

The chemistry of ions is not only of laboratory interest but very much a part of the practical world. Let us look at one aspect of the chemistry of ions as it affects our everyday lives, the chemistry of washing.

Soap is an ionic compound. The positive ion is usually the Na$^+$ ion, and the negative ion is a long hydrocarbon chain which terminates with a carboxyl group, $-\overset{\overset{O}{\parallel}}{C}-O^-$. A typical soap is represented by the formula

$$\left[H-\overset{H}{\underset{H}{C}}-\overset{H}{\underset{H}{C}}-\overset{H}{\underset{H}{C}}-\overset{H}{\underset{H}{C}}-\overset{H}{\underset{H}{C}}-\overset{H}{\underset{H}{C}}-\overset{H}{\underset{H}{C}}-\overset{H}{\underset{H}{C}}-\overset{H}{\underset{H}{C}}-\overset{H}{\underset{H}{C}}-\overset{H}{\underset{H}{C}}-\overset{H}{\underset{H}{C}}-\overset{H}{\underset{H}{C}}-\overset{H}{\underset{H}{C}}-\overset{H}{\underset{H}{C}}-\overset{H}{\underset{H}{C}}-\overset{H}{\underset{H}{C}}-\overset{O}{\underset{O^-}{\overset{\parallel}{C}}} \right] + Na^+$$

The negative ion is responsible for the cleansing action of soap, which is due to the different characteristics of the two ends of the ion. The hydrocarbon end is soluble in oil or grease, while the ionic end is soluble in water. Dirt is usually associated with oily materials, such as grease from food, skin oil, or lubricating oils. When the negative ions of soap are located where oil and water meet, they line up with the hydrocarbon end in the oil and the negative end in the water. Upon agitation, the oily material is dragged from the skin or clothes and is dispersed as tiny droplets in the water (see Fig. 9-8).

Natural water contains many dissolved materials. The ions present in water often include the following:

Na$^+$, K$^+$, Ca^{++}, Mg^{++}, NH$_4^+$, H$^+$, OH$^-$, NO$_3^-$, Cl$^-$, SO$_4^{--}$, and HCO$_3^-$

Of these, Ca^{++} and Mg^{++} are objectionable because they form an insoluble compound (precipitate) with soap. That is, these metallic ions do not stay in solution with the negative soap ion. The following equation represents this precipitation reaction:

$$Ca^{++} + 2CH_3(CH_2)_{16}COO^- \longrightarrow Ca[CH_3(CH_2)_{16}COO]_2$$

175 CHEMISTRY OF SOAP AND DETERGENTS

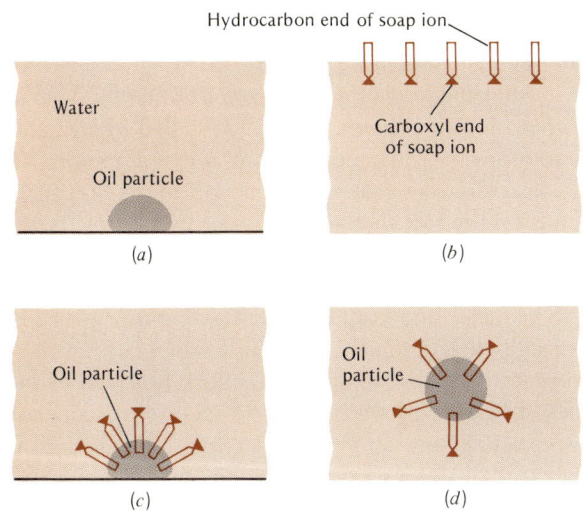

Figure 9-8 (a) Since water is a polar liquid and oil is a nonpolar liquid, water alone has no effect upon the oil particle adhering to the surface. (b) At the surface of the water the carboxyl end of the soap ion (represented by the triangle) is below the surface. The hydrocarbon end of the soap ion is above the surface. (c) At the water-oil interface the carboxyl end of the soap ion is in the water, and the hydrocarbon end is in the oil adhering to the surface. (d) Agitation results in the oil particle's being pulled from the surface into the body of the liquid.

This insoluble compound is responsible for "tattle-tale gray" and "bathtub ring." Soap cannot perform the cleansing action until all the Ca^{++} and Mg^{++} ions have been removed. Of course, if enough soap is used, these ions will be removed and the cleansing action takes place.

Water which contains Ca^{++} or Mg^{++} is called *hard water*, and removal of these ions is referred to as *softening* water. As we have seen, water can be softened with soap, but since soap is expensive, we remove the interfering ions with less expensive materials before soap is added. Water can be softened inexpensively with the carbonate and phosphate ions. The following equations are illustrative:

$$Ca^{++} + CO_3^{--} \longrightarrow CaCO_3$$
$$3Mg^{++} + 2PO_4^{3-} \longrightarrow Mg_3(PO_4)_2$$

In both cases the ions are used in the form of sodium compounds, Na_2CO_3 and Na_3PO_4. The sodium ion goes into solution; however, an increase in the concentration of Na^+ ion does not interfere with the cleansing action of soap. The net effect of these procedures of water softening is to replace Mg^{++} and Ca^{++} with Na^+.

Another method of softening water involves using insoluble sodium aluminosilicates. These compounds, called *zeolites*, consist of huge three-dimensional porous negative ions and sodium ions. (Review Chap. 6 for the structure of silicates.) The action of zeolites as water softeners depends on the fact that as hard water percolates through the channels of the huge negative ion, the Na^+ ions in zeolite are replaced by the Ca^{++} or Mg^{++} from the water. That is, the net effect here is the same as it was with Na_2CO_3 and Na_3PO_4: the Ca^{++} and Mg^{++} ions in the water are replaced by Na^+ ions.

In recent years soap has largely been replaced by synthetic detergents. These compounds are similar in structure to soap, having a long hydrocarbon chain and an ionic end group. They differ in the nature of the ionic end group. One type of ionic end group is the sulfonate groups, characterized by a sulfur-to-carbon bond in which the carbon atom is also a part of a benzene ring. A typical detergent is represented by the formula

$$\left[H-\underset{H}{\overset{H}{C}}-\underset{H}{\overset{H}{C}}-\underset{H}{\overset{H}{C}}-\underset{H}{\overset{H}{C}}-\underset{H}{\overset{H}{C}}-\underset{H}{\overset{H}{C}}-\underset{H}{\overset{H}{C}}-\underset{H}{\overset{H}{C}}-\underset{H}{\overset{H}{C}}-\underset{H}{\overset{H}{C}}-\underset{H}{\overset{H}{C}}-\underset{H}{\overset{H}{C}}-\bigcirc-\underset{O}{\overset{O}{S}}-O^- \right] + Na^+$$

The cleansing action of synthetic detergents is the same as that of soap, but detergents have an important advantage in that they are reasonably effective in hard water. Since they are even more effective in soft water, manufacturers add water softeners to their detergents. This is one of the reasons for adding phosphates.

A very distressing aspect of the widespread use of detergents is the part they play in water pollution. Foam caused by detergents is found in many of our waterways and even in our drinking water. Some detergents such as the one discussed above, are particularly harmful since, unlike soap, they are not degradable; that is, neither phosphate ions nor the alkylbenzene sulfonates undergo biochemical decomposition as most organic materials do. Furthermore, detergents are not removed by sewage treatment, and as a result they accumulate in our streams and lakes. Fortunately, detergents can be manufactured which are degradable. It is to be hoped that manufacturers will shift to this type of detergent.

Questions

1 Can you think of any reactions other than those mentioned in this chapter which take place spontaneously, that is, reactions which require merely that the reactants be brought into contact with each other?

QUESTIONS

2 How does free energy vary with changes in potential energy? How does it vary with changes in entropy?

3 What physical evidence can be cited to support a claim that the evaporation of a liquid does not simply increase entropy but also decreases potential energy?

4 In which situation is entropy the larger?
  a The dominoes at the start or end of the game
  b The ingredients of a cake before or after the batter is prepared for the oven
  c The pieces of a jigsaw before or after the puzzle is assembled
  d Milk immediately after milking or after sitting for several hours
  e Money in the cash register or in the collection plate at church
  f Billiard balls racked up before the game begins or after struck by cue ball
  g Dynamite before or after explosion
  h Race cars immediately before or after race commences

5 For each of the following changes tell whether potential energy is increased or decreased and whether entropy is decreased or increased:
  a The boiling of water   **b** The freezing of water
  c The burning of coal   **d** The leaking of a tire

6 $Ca(NO_3)_2$ is soluble in water, and $CaCO_3$ is insoluble in water. Explain in terms of forces between ions, between molecules, and between molecules and ions.

7 When NH_4NO_3 is added to water, the temperature of the water drops markedly. However, when $LiNO_3$ is added to water, the temperature increases considerably. Explain in terms of crystal energy and hydration energy.

8 Classify the following chemical changes which you have observed in everyday life as exothermic or endothermic:
  a Using a storage battery to start a car
  b Charging a storage battery
  c Burning a match

9 Criticize the following statements:
  a Dissolving NaCl is a physical process.
  b Freezing water is a physical process.
  c Water is a pure substance.

10 Describe what takes place when ionic crystals dissolve in water.

11 If all other factors are equal, which is more soluble? Explain your answer.
  a A solute with a high crystal energy or one with a low crystal energy
  b A solute with a high hydration energy or one with a low hydration energy

c A solute which dissolves with liberation of energy or one which absorbs energy upon dissolving

d A solute which has a great attraction for water molecules and hence increases the orderliness of water molecules upon dissolving or a solute which has very little attraction for water molecules and hence has little effect on the orderliness of water molecules upon dissolving

12 How many moles of silver nitrate are there in 3.00 liters of a 0.200 M solution? To how many grams of silver nitrate does this correspond?

13 What weight of $NaNO_3$ is needed to prepare 2.00×10^2 ml of a 3.00 M solution?

14 If the solubility of AgCl is 1.0×10^{-5} mole/liter:

a How many moles of Ag^+ ion are there in a liter of saturated solution?

b How many moles of Cl^- ion are there in a liter of saturated solution?

c What is the molarity of the saturated solution?

d How many grams of AgCl are there in a saturated solution?

e Should AgCl be classified as soluble or insoluble? Why?

15 Write the equation for the chemical reaction that takes place when members of each of the following pairs of aqueous solutions are mixed. If no reaction takes place, indicate by writing N.R. (You may wish to consult Appendix D for information on nomenclature.)

a $AgNO_3$ and Na_2S **b** NaCl and $Pb(NO_3)_2$
c NH_4Cl and $NaNO_3$ **d** NaOH and $AlCl_3$
e Na_3PO_4 and $Pb(NO_3)_2$ **f** K_2SO_4 and $Ba(NO_3)_2$
g Ammonium bromide and lead chlorate
h Sodium iodide and potassium sulfate
i Iron(III) chloride and sodium hydroxide
j Iron(II) chloride and potassium perchlorate

16 What is an advantage of synthetic detergents over soap? What is an advantage of soap over synthetic detergents?

17 In what ways are the structures of soap and synthetic detergents similar? How are they different?

18 Describe the cleansing action of soap.

19 Write an equation illustrating the softening of hard water.

CHAPTER 10   PROTON EXCHANGE (ACIDS AND BASES)

Various corrosive and noxious vapors are spewed into the air by industrial plants. Sulfur dioxide is a common constituent of these vapors. The effect of this gas on vegetation is shown in these two photographs. The first shows sulfur dioxide injury to tulip leaves. (R. H. Daines, Department of Plant Biology, Rutgers University.) The second shows the denuded hills surrounding a copper-smelting plant. The vegetation on these hills has been killed by the release of sulfur dioxide formed as a result of the processing of copper ores. (U.S. Department of Agriculture.)

CHAPTER 10   PROTON EXCHANGE (ACIDS AND BASES)

In Chap. 3 we discussed the formation of ions by the transfer of one or more electrons. Ions are also formed by the transfer of one or more protons, but we hasten to add that these protons come *only* from combined hydrogen atoms. The transferred proton, the nucleus of a hydrogen atom, is originally covalently bonded to another atom. This latter atom retains the bonding pair of electrons when the single proton of the hydrogen nucleus is transferred. Ammonium chloride is an example of a compound formed by a proton's being transferred from a hydrogen chloride molecule to an ammonia molecule:

$$H\!:\!\ddot{\underset{..}{Cl}}\!: \; + \; H\!:\!\ddot{\underset{H}{N}}\!:\!H \longrightarrow \left[H\!:\!\underset{H}{\overset{H^+}{\ddot{N}}}\!:\!H \; + \; :\!\ddot{\underset{..}{Cl}}\!:^- \right]$$

It is customary to refer to a proton donor as an *acid* and a proton acceptor as a *base*. Thus in the reaction above, hydrogen chloride is an acid and ammonia is a base.

What structural features must a molecule possess in order to act as a proton donor? Obviously it must contain at least one hydrogen atom. However, this alone is not sufficient to classify a compound as an acid. The methane molecule, CH_4, contains four combined hydrogen atoms, but rarely, if ever, does it function as a proton donor. This is true of hydrocarbons in general; hence they are not generally classified as acids. How does one distinguish acids from hydrogen-containing compounds which do not function as proton donors? The demarcation line is not a sharp one, and no general rule can be given to distinguish between the two groups of compounds. As so often is the case in chemistry, there is a gradual progression from one category to another. We saw in Chap. 4 that a group of elements known as the metalloids, which have both metal and nonmetal characteristics, occupies an area on the periodic table between the metals and nonmetals. Similarly we saw in Chap. 5 that there is a group of compounds containing bonds that are partially ionic and partially covalent. A similar phenomenon is encountered in distinguishing hydrogen-containing compounds which function as acids from hydrogen-containing compounds which do not function as acids. At one end of the spectrum we have such species as the hydrogen chloride molecule, which readily transfers its proton. It is classified as a *strong acid*. The hydrogen fluoride molecule transfers its proton less readily and is called a *weak acid*. The methane molecule never transfers a proton except under certain extreme conditions and is not ordinarily classified as an acid. There is no sharp line between weak acids and nonacids.

The structural feature common to all bases is an unshared pair of electrons. It is this unshared pair of electrons that makes a base able to accept a proton, forming a coordinate covalent bond. Like acids, bases differ in strength. That is, there is a difference in the ease with which they accept protons. The hydroxide ion, :Ö:H⁻, is a strong base and readily accepts protons. The ammonia molecule, H:N̈:H with H below, accepts a proton less readily and is classified as a weak base. Although the chloride ion has four unshared electron pairs, it does not accept a proton as readily as the ammonia molecule and hence is an even weaker base.

Some molecules and ions have structural features of both acids and bases and function as both. We indicated previously that pure water contains ions. These ions result from an acid-base reaction in which a proton is transferred from one water molecule to another, forming the hydroxide ion, :Ö:H⁻, and the hydronium ion, H:Ö:H⁺ with H below. In this reaction one water molecule acts as an acid. Another acts as a base. The extent of this reaction is very slight. About 1 water molecule in 554 million loses its proton, so that in 1 liter of water there will be 10^{-7} mole of hydronium ion and 10^{-7} mole of hydroxide ion.

Acids and Bases in Aqueous Solutions

Unlike ion-exchange reactions, acid-base reactions do not require a solvent. The reaction between ammonia and hydrogen chloride takes place simply upon mixing the two gases. However, since water is the medium in which many chemical reactions do take place, it is appropriate that special emphasis be placed upon acid-base reactions in this solvent.

We have indicated that the water molecule can act either as an acid or as a base. When hydrogen chloride is dissolved in water, water acts as a base. The following reaction takes place:

$$HCl + H_2O \longrightarrow H_3O^+ + Cl^-$$

Note that molecules react to form ions (see Fig. 10-1). This reaction and other reactions where nonionic substances react to form ions are referred to as *ionization* reactions. Another ionization reaction is the previously discussed reaction between ammonia and hydrogen chloride. It should be noted that dissolving ionic substances such as NaI in water is not an ionization reaction, since the ions existed before solution took place. The dissolving action simply permits the ions to move apart. This process is called *dissociation*. Dissolving NaI is not a chemical reaction, but dissolving hydrogen chloride is, since distinctly new species, H_3O^+ and Cl^-, are created.

The solution of hydrogen chloride in water has a hydronium concentration many times greater than the 10^{-7} M hydronium-ion concentration of water. If 0.100 mole of hydrogen chloride is dissolved in sufficient

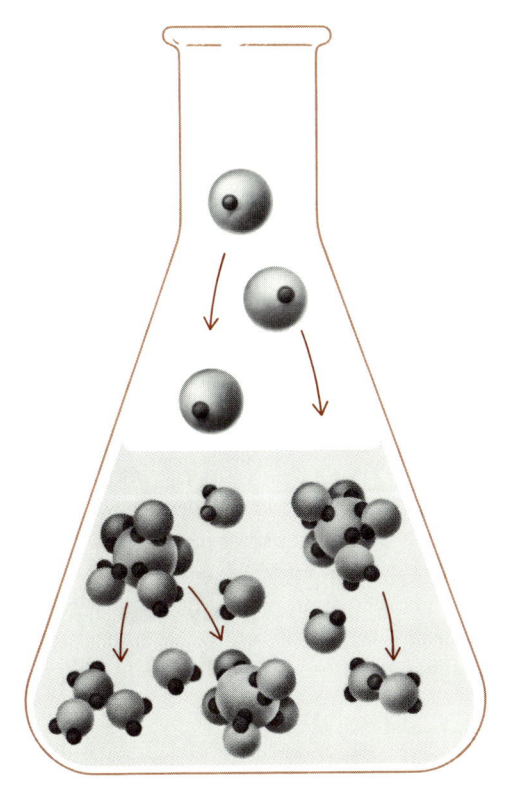

Figure 10-1 Polar water molecules pull HCl molecules apart to form hydronium ions and chloride ions.

water to produce 1.00 liter of solution, the hydronium-ion concentration will be 0.100 M, which is many times greater than the 10^{-7} M for hydronium ion of water. Solutions of hydrogen chloride and solutions of other substances in water which give a greater hydronium concentration than 10^{-7} M by ionization are known as *aqueous acids*. Other substances classified as aqueous acids when dissolved in water include HNO_3, H_2SO_4, HBr, HNO_2, H_3PO_4, and H_2S.

Aqueous acids are characterized by several distinctive properties. These include sour taste and the ability to change the color of certain vegetable dyes known as indicators. These acids also neutralize aqueous bases (see page 190) and in some cases react with carbonates and metals.

Aqueous bases are substances which either contain the OH^- ion or ionize in water to form the hydroxide ion so that the solution has a hydroxide concentration greater than 10^{-7} M. Sodium hydroxide is a solid which consists of Na^+ and OH^- ions. When it is dissolved, the ions dissociate, providing an OH^- concentration greater than 10^{-7} M. Ammonia forms an aqueous base by reacting with water to form NH_4^+ and OH^- ($NH_3 + H_2O \rightleftharpoons NH_4^+ + OH^-$). In this reaction, ammonia is a base and accepts a proton from water, which is an acid in this reaction. The aqueous base in this case is formed by ionization rather than by dissociation.

The Hydronium Ion versus the Hydrogen Ion

The hydrogen ion, H^+, is an exceptional ion because the nucleus is not screened by electrons. It consists only of a bare proton and exists independently only at such high energies as those provided in cathode-ray tubes. Certainly it does not exist in aqueous solutions. We have shown the proton as bonded to one water molecule in water and aqueous solutions. Because water molecules themselves are joined by hydrogen bonding, it seems likely that the proton is bonded to a variable number of water molecules in aqueous solutions. Probably hydrated protons are $H_5O_2^+$, $H_7O_3^+$, $H_9O_4^+$, etc. Because of the difficulty of representing the hydronium ion in terms of the correct number of water molecules bonded to the proton, some chemists prefer to use simply the H^+ in discussing aqueous acids. This viewpoint is in agreement with the custom of not showing the water of hydration associated with other ions. (Most ions are hydrated in solution.) However, in this book, we shall adhere to the prevailing custom of representing the hydronium ion by H_3O^+.

Acid-Base Equilibria

The examples of equilibria discussed up to this point have involved physical changes. We have learned that in a saturated solution the rate of dissolving the solute is equal to the rate of crystallizing the solute and the concentration of the solution thus remains constant. We have also learned that under certain conditions the rate of evaporation is equal to the rate of condensation and that the vapor pressure of the liquid remains constant.

Most acid-base reactions are examples of chemical equilibria. Chemical equilibria, like physical equilibria, do not go to completion. We have indicated that water undergoes autoionization to form hydronium ions and hydroxide ions. However, this reaction produces only minute concentrations of these ions. This does not mean that the reaction between water molecules proceeds until 10^{-7} mole of hydronium ion are formed per liter and then suddenly stops. The reaction continues indefinitely; but an opposing reaction between H_3O^+ and OH^- begins as soon as they are formed. That is, the reaction $H_3O^+ + OH^- \longrightarrow H_2O$ reduces the H_3O^+ and OH^- concentration. At equilibrium the rates of the opposing reactions are equal, and the concentrations of the H_3O^+ and OH^- ions remain constant.

For any system at equilibrium, the rates of opposing reactions are equal, and the concentrations of the components remain constant. That is, the driving force to form products is equal to the driving force to form reactants. Similarly, a balance has been reached between the conflicting requirements for minimum free energy, and the change in free energy is zero. This means that a state of balance between the opposing requirement for minimum energy content and maximum entropy has been reached.

A system remains at equilibrium until some condition is changed, causing the rate of one reaction to increase over the rate of the opposing reactions. When this is done, we say the equilibrium shifts. It is customary to represent chemical equilibrium by a double arrow:

$$H_2O + H_2O \rightleftharpoons H_3O^+ + OH^-$$

The equilibrium is said to be shifted to the left or right depending on which components of the equilibrium increase in concentration as a result of the shift.

One way to shift an equilibrium is to add one of the components of the equilibrium. For example, if the hydroxide-ion concentration is increased through the addition of sodium hydroxide crystals, the above equilibrium will react in such a way as to relieve the stress applied. That is, it will move to the left to reduce hydroxide-ion concentration. As the hydroxide-ion concentration is reduced, the rate of the reaction to the left decreases, and eventually the two opposing rates will become equal again and a new equilibrium will be established. When this new equilibrium is established, the hydroxide-ion concentration will be much larger than it was in water but smaller than it would have been if the reaction had not shifted upon the addition of sodium hydroxide crystals. The hydronium-ion concentration, of course, is much smaller than 10^{-7} M at the new equilibrium. However, it *never* becomes zero. A chemical equilibrium will not go to completion, no matter how great the stress applied.

Since the sodium ion is not a component of the equilibrium and does not react with any component of the equilibrium, its presence as a result of the addition of the NaOH crystals has no appreciable effect on the equilibrium. The concentration of sodium ion is the same before and after the shift has taken place.

Another method of shifting an equilibrium is to remove one of the components of the system. If we remove some hydronium ions, the reaction will shift to the right. One way of removing hydronium ions is to add an ion such as the nitrite ion, NO_2^-, which has a greater attraction for protons than water has. Since it is impossible to add the nitrite ion (or any other ion) by itself, we must select a salt containing it. Here we shall select sodium nitrite, $NaNO_2$. The sodium ion does not have any effect on the equilibrium.

When nitrite ions are dissolved in water, a second equilibrium is established. That is, we have two simultaneous equilibria. The equilibrium involving the nitrite ion ($NO_2^- + H_3O^+ \rightleftharpoons HNO_2 + H_2O$) applies a stress to the water equilibrium, causing it to shift to the right. When the new equilibrium is reached, the hydronium-ion concentration is smaller than it was in water but larger than it would have been if the equilibrium had not been shifted. The hydroxide-ion concentration at the new equilibrium is more than it was in water. Because the hydroxide-ion concentration in sodium nitrite solution is greater than it is in water, we say that the solution of sodium nitrite is basic.

A third method of shifting an equilibrium is to change the temperature. Reactions which proceed with the absorption of energy (endothermic) are shifted to increase the ratio of products to the reactions by an increase in temperature. Reactions which proceed with liberation of energy (exothermic) are shifted to decrease the ratio of products to reactants by an increase in temperature. That is, endothermic reactions are favored by an increase in temperature and exothermic reactions by a decrease in temperature. The reason is that at high temperatures the heat absorbed by an endothermic reaction is more readily acquired *from* the surroundings whereas at low temperatures the heat liberated by an exothermic reaction is more readily transferred *to* the surroundings.

Finally, some equilibria can be shifted by a change in pressures. Pressure changes can affect the relative proportions of products and reactants only if pressure change brings about a change in concentration. Since pressure does not ordinarily affect the volume of a liquid or solid, it has no effect on equilibria that involve reactants and products in these physical states. However, increases in pressure crowd molecules of gases closer together and raise their concentrations. Increases in pressure shift equilibria in the direction of the least number of molecules. For example, in the equilibrium $N_2 + 3H_2 \rightleftharpoons 2NH_3$, where all components are in the gaseous state, an increase in pressure causes the reaction to move to the right. In this reaction, four gas molecules are reacting to form two gas molecules. Since pressure and number of gas molecules are directly proportional, this decrease in the number of gas molecules relieves the stress of increased pressure.

Polyprotic Acids Some acids, called polyprotic acids, are able to transfer more than one proton per molecule. Sulfuric acid is an example. When it is added to water, the first proton is lost from all the molecules to form hydronium ions and hydrogen sulfate ions, HSO_4^-. That is, the reaction $H_2SO_4 + H_2O \longrightarrow H_3O^+ + HSO_4^-$ goes to completion. Further ionization of the hydrogen sulfate occurs. Since the hydrogen sulfate ion is a weak acid, the reaction does not go to completion and an equilibrium, $HSO_4^- + H_2O \rightleftharpoons H_3O^+ + SO_4^{--}$, is reached when only a small portion of the HSO_4^- ions have ionized.

IONIC ACIDS

Phosphoric acid, H_3PO_4, another polyprotic acid, ionizes in three stages. The first ionization reaches a state of equilibrium when only a small portion of the first protons has been transferred to water molecules. An even smaller fraction of the second protons is lost to water molecules when equilibrium is reached. The transfer of the third protons takes place to an even lesser extent than the second. Thus phosphoric acid is a weak acid in all three stages of ionization.

The ionization of phosphoric acid is represented by the following:

$$H_3PO_4 + H_2O \rightleftharpoons H_3O^+ + H_2PO_4^-$$
$$H_2PO_4^- + H_2O \rightleftharpoons H_3O^+ + HPO_4^{2-}$$
$$HPO_4^{2-} + H_2O \rightleftharpoons H_3O^+ + PO_4^{3-}$$

Other polyprotic acids include carbonic acid, H_2CO_3; sulfurous acid, H_2SO_3; and oxalic acid, $H_2C_2O_4$. All are weak acids in both stages of ionization.

One should not conclude that just because an acid contains more than one hydrogen atom per molecule it is polyprotic. For example, acetic acid

Acetic acid

is monoprotic. Only the proton bonded to the oxygen is transferable. In general, hydrogen atoms bonded to carbon atoms cannot be transferred.

Ionic Acids   We have seen that negative ions such as the HSO_4^- and $H_2PO_4^{2-}$ ions can act as acids, and we discussed these ions as intermediates in the ionization of polyprotic acids. These acid ions can also be obtained in solution by dissolving the compounds $NaHSO_4$ and NaH_2PO_4. Both compounds produce solutions with a hydronium-ion concentration greater than 10^{-7} mole/liter. That is, solutions of these salts are acid.

Positive ions can also act as acids. Ammonium chloride dissolved in water produces an acid solution because of the equilibrium

$$NH_4^+ + H_2O \rightleftharpoons H_3O^+ + NH_3$$

Numerous hydrated metal ions also produce acid solutions. For example, a solution of iron(III) chloride produces an acid solution because of the equilibrium

$$Fe(H_2O)_6^{3+} + H_2O \rightleftharpoons H_3O^+ + Fe(OH)(H_2O)_5^{2+}$$

188 PROTON EXCHANGE (ACIDS AND BASES)

Strength of Acids

In every acid-base reaction another acid-base pair is produced. For example, in the reaction $F^- + H_2O \rightleftharpoons HF + OH^-$ the water molecule is the acid and the fluoride ion is the base when the reaction proceeds from left to right, but when the reaction proceeds from right to left, the hydrogen fluoride molecule is the acid and the hydroxide ion is the base.

Similarly, in the reaction $Cu(H_2O)_4^{++} + H_2O \rightleftharpoons Cu(OH)(H_2O)_3^+ + H_3O^+$, the $Cu(H_2O)_4^{++}$ ion is the acid and the water molecule is the base when the reaction proceeds from left to right. The hydronium ion is the acid and the $Cu(OH)(H_2O)_3^+$ is the base when the reaction proceeds from right to left.

It is possible to determine experimentally the relative strengths of acids and bases. These experimental data are presented in Table 10-1. The acids are arranged in order of decreasing strength and the bases in order of increasing strength. The term *conjugate base* is used to refer to the base

TABLE 10-1 Relative Strengths of Some Acids and Their Conjugate Bases

	Acid	Conjugate Base	
Decreasing strength of acid ↓	$HClO_4$	ClO_4^-	Increasing strength of base ↓
	HI	I^-	
	HBr	Br^-	
	H_2SO_4	HSO_4^-	
	HCl	Cl^-	
	HNO_3	NO_3^-	
	$HClO_3$	ClO_3^-	
	H_3O^+	H_2O	
	H_2SO_3	HSO_3^-	
	HSO_4^-	SO_4^{--}	
	$HClO_2$	ClO_2^-	
	H_3PO_4	$H_2PO_4^-$	
	$Fe(H_2O)_6^{3+}$	$Fe(OH)(H_2O)_5^{++}$	
	H_3AsO_4	$H_2AsO_3^-$	
	H_2Te	HTe^-	
	HF	F^-	
	H_2Se	HSe^-	
	$HC_2H_3O_2$	$C_2H_3O_2^-$	
	$Al(H_2O)_6^{3+}$	$Al(OH)(H_2O)_5^{++}$	
	H_2CO_3	HCO_3^-	
	H_2S	HS^-	
	$H_2PO_4^-$	HPO_4^{--}	
	HSO_3^-	SO_3^{--}	
	HClO	ClO^-	
	NH_4^+	NH_3	
	HCO_3^-	CO_3^{--}	
	HS^-	S^{--}	
	H_2O	OH^-	
	NH_3	NH_2^-	
	CH_4	CH_3^-	

STRENGTH OF ACIDS

formed as a result of the proton transfer by the acid. Similarly, the term *conjugate acid* is used to refer to the acid formed as a result of a proton gained by a base. Since the strength of the bond to the proton increases as one goes down in this table, the reactions of acids with bases below them in the table tend to go to completion. However, there is little or no tendency for acids to donate protons to bases above them in the table.

The rankings in this table illustrate a number of generalizations regarding the strength of acids and bases. The positions of H_2S relative to HS^- and of H_2SO_3 relative to HSO_3^- illustrate the generalization that a polyprotic acid ionizes to a greater extent in the first ionization than in the second. This is easy to understand when we realize that in the second ionization the proton is released from an ion, SO_3^{--}, with twice the charge of the ion, HSO_3^-, from which the first proton escapes.

The ranking of HF, H_2O, NH_3, and CH_4 exemplifies the generalization that within a period of the periodic table the acidity of a binary-hydrogen compound decreases from right to left. This is explained by the decreasing electronegativity of the atom to which the proton is bonded and hence the decreasing polarity of the bond holding the proton.

Considering what has been said in the previous paragraph, one might be surprised at the positions of HI, HBr, HCl, and HF. The strength of an acid increases as one goes down within a group of the periodic table. That is, the strength of the acid increases as the electronegativity of the atom to which the proton is bonded decreases. This anomaly is explained by the increased size of the atom with increasing atomic number in a group. This increased size results in greater surface area over which the negative charge is distributed and a weakening in the force of attraction for the proton.

The strength of acids in which hydrogen atoms are covalently bonded to oxygen atoms in turn covalently bonded to a third element (*oxyacids*) increases with the number of oxygen atoms per molecule. The locations of $HClO_4$, $HClO_3$, $HClO_2$, and $HClO$ illustrate this premise. Here the highly electronegative oxygen atoms draw electrons away from the O—H bond, making it easier to remove the proton.

The strength of oxyacids within a period increases from left to right, as illustrated by the positions of H_3PO_4, H_2SO_4, and $HClO_4$. This ranking is accounted for by the increased electronegativity of the central atom, resulting in a greater pull on the electrons of the O—H bond and making it easier for the proton to be released.

The decrease in the strength of the oxyacids as one goes down a periodic group is exemplified by the positions of HNO_3, H_3PO_4, and H_3AsO_4. Again this ranking is accounted for by changes in electronegativity. The decrease in electronegativity as one goes down a group results in a smaller pull on the electrons of the O—H bond and hence the proton is released less readily.

Neutralization When a hydroxide is added to an aqueous acid in such quantity that the number of moles of OH^- is exactly the same as the number of moles of monoprotic acid or exactly twice the number of moles of diprotic acid or exactly three times the number of moles of a triprotic acid, etc., we say the acid has been neutralized. This process consists essentially of the removal of protons from the acid molecules to form water molecules with the hydroxide ions. The addition of sodium hydroxide, NaOH, to hydrochloric acid so that the number of moles of hydroxide ion equals the number of moles of hydrochloric acid is an example of neutralization. This reaction is represented by the equation $H_3O^+ + OH^- \longrightarrow 2H_2O$. The ions of sodium chloride remain in solution. This compound can be procured if the water is evaporated.

The neutralization of hydrochloric acid by sodium hydroxide results in a solution with both hydronium-ion and hydroxide-ion concentrations equal to 10^{-7} mole/liter, the same as in water. Such a solution is said to be *neutral*. This is true because neither the sodium ion nor the chloride ion affects the water equilibrium. However, if acetic acid, $HC_2H_3O_2$, is neutralized by sodium hydroxide, the resulting solution is *not* neutral. This is because the acetate ion, $C_2H_3O_2^-$, which is formed by neutralization, does affect the water equilibrium. The equilibrium $C_2H_3O_2^- + H_3O^+ \rightleftharpoons HC_2H_3O_2 + H_2O$ is established, which causes the water equilibrium $H_2O + H_2O \rightleftharpoons H_3O^+ + OH^-$, to move to the right, forming a basic solution. Similarly, if NH_3 is used to neutralize hydrochloric acid, an acid solution results. Here the ammonium ion, NH_4^+, produced by the neutralization process, disturbs the water equilibrium by producing hydronium ions: $NH_4^+ + H_2O \rightleftharpoons H_3O^+ + NH_3$. It is generally true that the neutralization of strong acids (above the hydronium ion in Table 10-1) by hydroxide ions results in neutral solutions. The neutralization of weak acids (below the hydronium ion in Table 10-1) by the hydroxide ion results in basic solutions, and the neutralization of strong acids by weak bases such as aqueous ammonia, results in acid solutions.

EVERYDAY INSTANCES OF ACID-BASE REACTIONS

Our discussion of acids and bases up to this point has been related to theoretical concepts. However, acids and bases and acid-base reactions are very much a part of our everyday lives. Oxyacids are readily prepared by dissolving oxides of nonmetals in water. Since carbon dioxide, CO_2, and water are very prevalent in our environment, carbonic acid, H_2CO_3, and its ions, HCO_3^- and CO_3^{--}, are often encountered. The following equations show the relations between these species:

$$CO_2 + H_2O \rightleftharpoons H_2CO_3$$
$$H_2CO_3 + H_2O \rightleftharpoons H_3O^+ + HCO_3^-$$
$$HCO_3^- + H_2O \rightleftharpoons H_3O^+ + CO_3^{--}$$

NEUTRALIZATION

We have indicated that aqueous acids have a sour taste. Carbonic acid, a weak acid, is used to impart a characteristic taste to soft drinks. The effect of carbonic acid on the taste of carbonated beverages can be experienced by leaving a bottle of carbonated beverage uncapped for about a half an hour at room temperature. The solubility of a gas in liquid is proportional to the pressure. Carbon dioxide was forced into the beverage under pressure and tightly capped to prevent the gas from escaping. As soon as the cap is removed, carbon dioxide escapes from the solution and bubbles of the gas form. We say it effervesces. Also the solubility of a gas decreases with an increase in temperature. Thus after a half an hour, most of the H_2CO_3 will have decomposed and the CO_2 escaped, and the beverage will be noticeably sweeter.

The HCO_3^- ion is both an acid and a base. However, it is a stronger base than acid. As a result, aqueous solutions of salts, such as $NaHCO_3$, are mildly basic. Because it is such a mild base, $NaHCO_3$ is used as an antacid; that is, it is taken internally to relieve hyperacidity in the stomach.

An important use of the bicarbonate ion is as an aerating agent. Because $NaHCO_3$ is marketed for this purpose, it is often called baking soda. The aerating action of the HCO_3^- ion results from its reaction with the hydronium ion to form unstable H_2CO_3, which subsequently decomposes to liberate CO_2. The gas bubbles formed by the CO_2 cause dough to rise in baking, giving it the desired lightness and porosity. The hydronium ions must be provided by adding some acidic material such as sour milk, which contains lactic acid. Mixtures which contain both $NaHCO_3$ and a compound which forms hydronium ions in aqueous solution are called baking powders. The compounds $Ca(H_2PO_4)_2$ and $NaAl(SO_4)_2 \cdot 12H_2O$ are commonly used for this purpose. Starch is also added to baking powders to help keep the mixture dry. No reaction takes place as long as the mixture stays dry. When water is added, the acidic component reacts with water to form hydronium ions:

$$H_2PO_4^- + H_2O \rightleftharpoons H_3O^+ + HPO_4^{--}$$
$$Al(H_2O)_6^{3+} + H_2O \rightleftharpoons Al(OH)(H_2O)_5^{++} + H_3O^+$$

Subsequently the H_3O^+ reacts with the HCO_3^- from baking soda to form H_2CO_3:

$$H_3O^+ + HCO_3^- \rightleftharpoons H_2CO_3 + H_2O$$

The carbonic acid immediately decomposes, giving CO_2:

$$H_2CO_3 \longrightarrow H_2O + CO_2$$

Another important aspect of the chemistry of the H_2CO_3-HCO_3^- system is its buffer action in blood. A buffer is a solution which maintains a nearly constant H_3O^+ concentration upon the addition of small amounts of H_3O^+ or OH^-. Usually buffers contain a weak acid and the conjugate base of the weak acid. If H_3O^+ enters blood, the HCO_3^- reacts with it:

$$H_3O^+ + HCO_3^- \rightleftharpoons H_2CO_3 + H_2O$$

And if OH^- enters blood, H_2CO_3 reacts with it:

$$OH^- + H_2CO_3 \rightleftharpoons HCO_3^- + H_2O$$

Although a buffer is never 100 percent efficient, the variation in H_3O^+ concentration is kept within rather narrow limits. Blood contains other buffers in addition to the H_2CO_3-HCO_3^- system. The combined action of these buffers maintains the hydronium-ion concentration in blood between 4.5×10^{-8} and 3.5×10^{-8} mole/liter. Thus blood is very slightly basic.

Acetic acid, $HC_2H_3O_2$, is another common household acid. Vinegar, which is about 4 percent acetic acid, is often used to impart an acid (sour) taste. Acetic acid, like the acid components of baking powder, is a stronger acid than carbonic acid. This is readily demonstrated in the kitchen by adding vinegar to baking soda and observing the bubbles of CO_2.

Anyone who has lived in a large industrial city afflicted by smog has unhappily encountered sulfuric acid, H_2SO_4. Smog develops when bright sunlight acts upon a mixture of gases from automobile exhaust and industrial plants trapped in a geographical region. Sulfur dioxide is among the gases expelled by industrial plants using coal or other sulfur-containing fuels. This gas subsequently reacts with oxygen to form sulfur trioxide, which dissolves in water to form sulfuric acid. The presence of SO_2 and H_2SO_4 in smog presents a serious health problem. In addition, sulfuric acid attacks building material containing carbonates, such as marble and limestone. Again the reaction is one in which carbonic acid is formed:

$$2H_3O^+ + CO_3^{--} \rightleftharpoons H_2CO_3 + 2H_2O$$

A partial answer to the problem of air pollution by sulfur dioxide is to require industrial plants to pass waste gases through ammonia solutions. This procedure removes more than 90 percent of the SO_2 by forming NH_4HSO_3. The formation of NH_4HSO_3 is represented by the following equations:

$$H_2O + SO_2 \rightleftharpoons H_2SO_3$$
$$H_2SO_3 + NH_3 \rightleftharpoons NH_4^+ + HSO_3^-$$

QUESTIONS

The compound NH_4HSO_3 can be recovered from the solution by evaporation. Not only does this procedure remove most of the SO_2, but it also produces a commercially valuable product. The textile industry uses NH_4HSO_3 in the production of nylon. This simple procedure is now used by a number of industrial plants in Holland.

Questions

1. What is the difference between the members of each of the following pairs of terms?
   a Hydrogen chloride and hydrochloric acid
   b Acid and base
   c Ionization and dissociation
   d Hydrogen ion and hydronium ion
   e Strong acid and weak acid
   f Aqueous acid and aqueous base
   g Baking soda and baking powder
2. Write a balanced chemical equation to show the reaction by which the ions of water are produced.
3. What role does water play in the ionization of hydrogen chloride in hydrochloric acid?
4. What is the concentration of H_3O^+ in 1.00 liter of solution containing 1.00 mole of HCl? What is the Cl^- concentration?
5. Write equations showing how aqueous acids are formed from the following anhydrous compounds.
   a HNO_3 b HBr c H_2SO_4
6. List the properties of aqueous acids.
7. Contrast NaOH and NH_3 according to the methods by which they increase the OH^- concentration upon dissolving in water.
8. Define chemical equilibrium in terms of:
   a Free energy
   b Quantities of products and reactants
   c Speed of reaction
   d Energy content and entropy
9. What is meant by a shift in an equilibrium? Use an example to explain.
10. What are three ways of changing an equilibrium?
11. Consider the equilibrium $NH_4^+ + H_2O \rightleftharpoons H_3O^+ + NH_3$.
    a What effect will the addition of NH_4Cl have on the H_3O^+ concentration?
    b What effect will the addition of HCl have on the NH_4^+ concentration?
    c What effect will the addition of NaCl have on the H_3O^+ concentration?
    d What effect will the addition of NaOH have on the NH_4^+ concentration?

194 PROTON EXCHANGE (ACIDS AND BASES)

  e What effect will the addition of NH_4Cl have on the NH_3 concentration?

12 How is the equilibrium $N_2 + 3H_2 \rightleftharpoons 2NH_3$ affected by the following changes? The reaction is exothermic to the right.
 a Pressure is decreased.   **b** Nitrogen is added.
 c Ammonia is added.   **d** Hydrogen is removed.
 e Ammonia is removed.   **f** Temperature is increased.

13 Write formulas for an acid in each of the following groups.
 a Polyprotic acid   **b** Strong acid
 c Weak acid   **d** Ionic acid

14 Give an example of each of the following structural units which will react with water to give an acid solution:
 a A molecule   **b** A negative ion   **c** A positive ion

15 List all the ions and molecules present in an aqueous solution of each of the following:
 a HCl   **b** H_2SO_4
 c H_2S **d** NH_3

16 List the acids in each group in order of decreasing strength and explain your ranking in terms of the structure of the acid:
 a $H_2PO_4^-$, H_3PO_4, HPO_4^{--} **b** SiH_4, HCl, PH_3, H_2S
 c H_2Te, H_2O, H_2S, H_2Se **d** H_3PO_3, H_3PO_2, H_3PO_4
 e H_2SO_4, H_4SiO_4, H_3PO_4 **f** H_3PO_4, H_3AsO_4, HNO_3

17 Show that 10 ml of 1.0 M HCl will neutralize 1.00 ml of 10 M NaOH.

18 Write the equations for the ionization of each of the following:
 a HCl   **b** $HC_2H_3O_2$
 c HNO_2 **d** H_2SO_4 (two equations)

19 Identify by formula the conjugate base of the following acids:
 a $HC_2H_3O_2$ **b** H_2S **c** HSO_3^-
 d NH_4^+ **e** HCl   **f** $Cu(H_2O)_4^{++}$
 g H_2O

20 Identify by formula the conjugate acid of each of the following bases:
 a S^{--} **b** H_2O **c** OH^-
 d NH_3 **e** NO_2^- **f** SO_4^{--}
 g HCO_3^-

21 Consult Table 10-1 and predict in each case whether equilibrium favors products or reactants:
 a $H_3PO_4 + F^- \rightleftharpoons HF + H_2PO_4^-$
 b $H_2SO_4 + ClO_2^- \rightleftharpoons HClO_2 + HSO_4^-$
 c $H_2S + C_2H_3O_2^- \rightleftharpoons HC_2H_3O_2 + HS^-$
 d $HClO + SO_4^{--} \rightleftharpoons HSO_4^- + ClO^-$
 e $NH_4^+ + F^- \rightleftharpoons HF + NH_3$
 f $HF + SO_4^{--} \rightleftharpoons HSO_4^- + F^-$

22 Write the fundamental equation for the neutralization reaction.
23 Classify the following as acid, base, both acid and base, or neither acid nor base:
   a Na^+ b F^- c H_2O
   d NH_4^+ e HSO_4^- f S^{--}
24 Predict whether solutions of each of the following substances will have a hydronium-ion concentration equal to, more than, or less than 10^{-7} M:
   a NaCl b NaOH c KNO_2
   d NH_4Cl e $NaC_2H_3O_2$ f Na_2S
   g HCl h $NaHSO_3$ i NaH_2PO_4
25 Criticize the following statements:
   a Acetic acid, $HC_2H_3O_2$, solution is neutralized when an equal number of moles of OH^- ion have been added.
   b The formula for the hydronium ion is H_3O^+.
26 Explain or account for the following statements:
   a A solution with a high OH^- concentration *must* have a low H_3O^+ concentration.
   b A solution of $NaNO_2$ is basic.
   c A solution of NH_4Cl is acid.
   d HI is a stronger acid than HF.
   e $HClO_4$ is a stronger acid than HClO.
   f H_2SO_3 is a stronger acid than HSO_3^-.
   g HF is a stronger acid than H_2O.
   h H_2SO_4 is a stronger acid than H_3PO_4.
   i Not all hydrogen compounds are classified as acids.
27 Classify the following as aqueous acids, aqueous bases, or neutral solutions:
   a A solution in which H_3O^+ concentration is 10^{-7}
   b A solution in which H_3O^+ concentration is 10^{-8}
   c A solution in which H_3O^+ concentration is 10^{-6}
   d Water
   e NaCl solution
   f $NaNO_2$ solution
   g $NaC_2H_3O_2$ solution
   h NH_4Cl solution
28 Oven cleaners are essentially sodium hydroxide (lye), which is a strong base and extremely caustic. After cleaning an oven, a housewife wishes to be sure that all the base has been removed. What common household product would you suggest to neutralize any excess sodium hydroxide?
29 Soap is a salt of a weak acid and sodium hydroxide. Would a solution of soap be basic, acid, or neutral? Why?

CHAPTER 11 ELECTRON EXCHANGE: OXIDATION AND REDUCTION

The development of the modern automobile is one of the great technological achievements of this century. However, as so often happens with technological advancements, too little attention has been paid to side effects. As a result, our major population centers now suffer from air pollution to an intolerable degree. (Planned Parenthood and World-Population Foundation.)

CHAPTER 11   ELECTRON EXCHANGE: OXIDATION AND REDUCTION

When we discussed compound formation by the transfer of electrons in Chap. 3, the discussion was limited to reactions between isolated atoms. However, except for rare gases, isolated atoms are very unusual and the monatomic state an extremely artificial one. Metals exist in crystals consisting of ions surrounded by a cloud of electrons, and nonmetals exist in molecular form, as described in Chaps. 6 and 8. Thus, chemical reactions by the transfer of electrons are in fact considerably more complicated than those described in Chap. 3.

Reactions in which electrons are exchanged are referred to as *oxidation-reduction reactions*. The reactant which donates electrons is called the reducing agent, and the reactant accepting electrons is called the oxidizing agent. The reactant which accepts electrons (oxidizing agent) is said to be reduced, and the reactant which donates electrons (reducing agent) is said to be oxidized.

Reactions between Metal Crystals and Positively Charged Ions

Most metal crystals are good electron donors (reducing agents), and many positively charged ions are good electron acceptors (oxidizing agents). For example, when zinc crystals are placed in hydrochloric acid, electrons are transferred from the zinc crystals to H_3O^+ ions, and a vigorous evolution of hydrogen ensues. This reaction is represented by the equation

$$Zn + 2H_3O^+ \longrightarrow Zn^{++} + H_2 + 2H_2O$$

It should be remembered that Zn in this equation stands for zinc crystals and not isolated Zn atoms.

Zinc will reduce other positively charged ions. When zinc is placed in an aqueous solution containing copper ions, copper crystals and zinc ions are produced:

$$Zn + Cu^{++} \longrightarrow Zn^{++} + Cu$$

When zinc is placed in an aqueous solution of magnesium ions, no reaction takes place. However, when magnesium is placed in a solution containing zinc ions, the following reaction takes place:

$$Mg + Zn^{++} \longrightarrow Mg^{++} + Zn$$

Extended experimentation of this sort leads to a comparison of metals on the basis of their ability to reduce hydronium ions and other positively charged ions. Table 11-1 lists common metals in order of the decreasing ease with which they lose electrons. Thus potassium, at the top of the table, is the strongest reducing agent in the table, and the noble metals—silver, platinum, and gold—are the weakest reducing agents.

TABLE 11-1 Comparison of Metals as Reducing Agents

Potassium
Sodium
Magnesium
Aluminum
Zinc
Iron
Cobalt
Nickel
Tin
Lead
[Hydrogen]
Copper
Mercury
Silver
Platinum
Gold

Every metal in this series will reduce ions of the metals below it but not the ions of the metals above it. Thus lead will reduce hydronium ions but not tin ions. Similarly, copper will reduce mercury ions but not hydronium ions. If the ions of these metals are considered, the ions of those at the bottom of the table are the strongest oxidizing agents and the ions of the metals at the top of the table are the weakest oxidizing agents. Thus, the potassium ion is the weakest oxidizing agent, and the gold ion is the strongest oxidizing agent.

Electro-chemistry

When zinc crystals are placed in a solution of copper ions, electrons are transferred directly from the zinc to the copper ions. This is an exothermic reaction. That is, the energy content of the products is less than that of the reactants. When the zinc crystals and copper ions are in direct contact, this energy is released as heat. It is possible to carry out this reaction without having the reactants in physical contact. In this situation, the electrons are not transferred from the zinc crystals directly to the copper ions but pass through an external path such as a copper wire. We have learned that a movement of electrons through a metal conductor is an electric current. Thus, in this case, some of the energy is released as electric energy rather than heat energy.

A picture of the apparatus used in carrying out the reaction between zinc and copper ions without physical contact is shown in Fig. 11-1. In one compartment copper is immersed in a solution containing copper ions. The other compartment contains zinc immersed in a solution containing zinc ions. The two compartments are separated by a porous divider which permits contact between the two solutions but prevents their mixing or diffusion. When the two pieces of metal (*electrodes*) are connected by a copper wire, zinc goes into solution as zinc ions (Zn $\longrightarrow$ Zn^{++} + 2e$^-$) and copper ions are deposited on the copper (Cu^{++} + 2e$^-$ $\longrightarrow$ Cu). That is, the oxidation-reduction reaction (Zn + Cu^{++} $\longrightarrow$ Zn^{++} + Cu) is divided into two half-reactions, with oxidation taking place in one compartment and reduction taking place in the other.

The reaction proceeds only if the solutions in the two compartments are kept electrically neutral. In the left compartment positively charged ions are formed, and in the right compartment positively charged ions are removed. It is obvious that migration of ions must be provided for. This is the purpose of the porous dividers. Positively charged ions migrate through the divider from left to the right, and negatively charged ions migrate from right to left, keeping both solutions electrically neutral.

The apparatus that we have described is called a *voltaic cell*. The electrode where reduction takes place (in this case the copper electrode) is called the *cathode*. The electrode where oxidation takes place is called the *anode*. A voltaic cell can be used to perform useful work. For example, a small light bulb placed in the circuit as shown in Fig. 11-1 will convert electric energy to light.

A voltaic cell will operate until one or both of the reactants is exhausted. For the cell just described, electric energy will be produced until either the copper ions or the zinc metal is completely consumed. The

Figure 11-1 The zinc-copper voltaic cell.

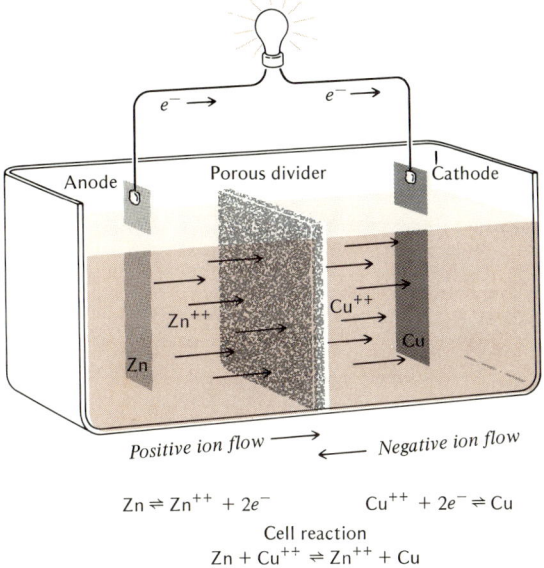

$$Zn \rightleftharpoons Zn^{++} + 2e^- \qquad Cu^{++} + 2e^- \rightleftharpoons Cu$$
$$\text{Cell reaction}$$
$$Zn + Cu^{++} \rightleftharpoons Zn^{++} + Cu$$

voltage, or intensity of the electric energy produced by a voltaic cell, depends on the nature of the electrodes, the concentration of the solutions in which they are immersed, and the temperature. The greater the difference between the two metals in reducing strength, that is, the farther apart they are in the series in Table 11-1, the greater the voltage of the cell. For example, a zinc-gold cell would normally produce a greater voltage than the zinc-copper cell. Electrodes are always immersed in a solution of their own ions. The highest voltage is produced when the anode is immersed in a dilute solution and the cathode in a concentrated solution.

Voltaic cells are electrochemical cells which utilize chemical reactions to produce electric energy. Another type, *electrolytic cells*, performs the opposite function. That is, they use electric energy to induce a chemical reaction. This process is called *electrolysis*. The essentials of an electolytic cell include two electrodes immersed in an *electrolyte* (a molten ionic compound or an aqueous solution containing ions), a source of direct current, such as a voltaic cell, and a copper wire to provide an external connection between the two electrodes.

In the typical electrolytic cell shown in Fig. 11-2 the electrolyte is melted sodium chloride. The electrodes are made from graphite, platinum, or some other chemically inactive conducting substance. When the two chemically inert electrodes are connected by a copper wire through a battery (voltaic cell), electrons are taken from one electrode, making it positive (anode), and deposited on the other electrode, making it negative (cathode). That is, the battery acts as an electron pump, removing electrons from one electrode and adding them to the other electrode.

The negative ions (Cl$^-$) are attracted to the positive electrode (anode),

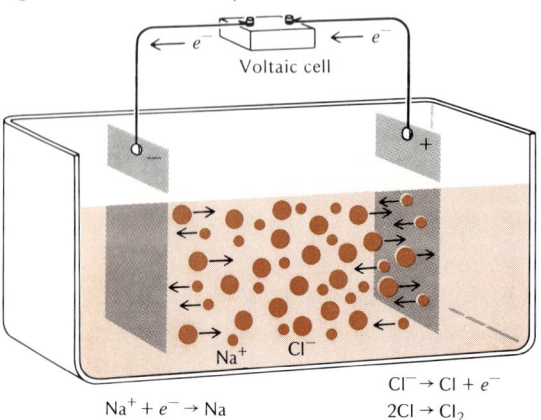

Figure 11-2 An electrolytic cell.

$Na^+ + e^- \rightarrow Na$

$Cl^- \rightarrow Cl + e^-$
$2Cl \rightarrow Cl_2$

and the positive ions (Na⁺) are attracted to the negative electrode (cathode). Upon reaching the anode, the chloride ions deposit electrons, forming neutral atoms. These neutral atoms immediately combine to form chlorine molecules:

$$:\ddot{\underset{..}{Cl}}:^- \longrightarrow :\ddot{\underset{..}{Cl}}\cdot + e^-$$

$$2:\ddot{\underset{..}{Cl}}\cdot \longrightarrow :\ddot{\underset{..}{Cl}}:\ddot{\underset{..}{Cl}}:$$

At the cathode, sodium ions pick up electrons and become sodium atoms:

$$Na^+ + e^- \longrightarrow Na$$

As in the voltaic cell, oxidation takes place at the anode and reduction at the cathode. The operation of the cell brings about the decomposition of sodium chloride into sodium and chlorine:

$$2NaCl \longrightarrow 2Na + Cl_2$$

It should be noted that sodium chloride crystals cannot be used as an electrolyte because in an ionic crystal the ions are held in a relatively fixed position. In order for an ionic substance to be an effective electrolyte, the ions must be relatively free to move independently. That is, the crystal structure must be broken down. This is accomplished by melting the crystal or dissolving it in a polar solvent such as water.

APPLICATION OF ELECTROCHEMISTRY

The voltaic cell which we have discussed is of theoretical importance only. However, there are a number of voltaic cells which are of considerable commercial importance. Among these is the so-called *dry cell* (the name is inappropriate since this cell does contain water), which is used to operate flashlights, portable radios, and other appliances.

This cell consists of a carbon rod surrounded by a paste of MnO_2, NH_4Cl, $ZnCl_2$, some inert material for a filler, and water. All this material is encased in a zinc container which serves as the anode (see Fig. 11-3). The anode reaction is $Zn \longrightarrow Zn^{++} + 2e^-$.

The carbon rod serves as the cathode. However, it is inert and does not take part in the reaction of the cell. The cathode reaction takes place on the surface of the carbon rod and is essentially one of reduction of the H_3O^+:

$$2e^- + 2H_3O^+ \longrightarrow H_2 + 2H_2O$$

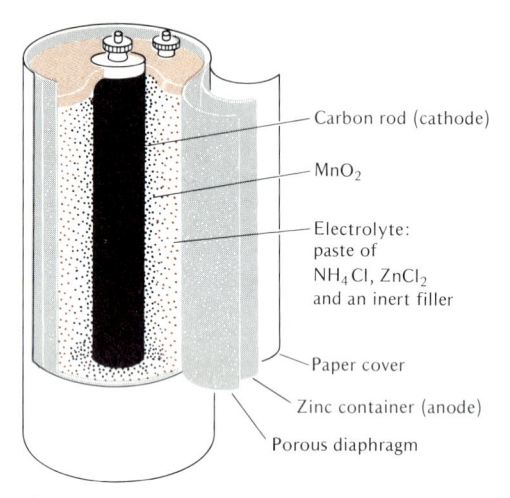

204 ELECTRON EXCHANGE: OXIDATION AND REDUCTION

[Diagram of dry cell with labels: Carbon rod (cathode); MnO$_2$; Electrolyte: paste of NH$_4$Cl, ZnCl$_2$ and an inert filler; Paper cover; Zinc container (anode); Porous diaphragm]

Figure 11-3 The dry cell.

The hydronium ion comes from the ionic acid, NH$_4^+$:

$$NH_4^+ + H_2O \rightleftharpoons H_3O^+ + NH_3$$

Both gases produced from the cell's operation must be removed to prevent the cell from bursting. The hydrogen is removed by a reaction with MnO$_2$ to form water. The ammonia is removed by forming coordinate covalent bonds with the Zn^{++} ion.

Theoretically it should be possible to reverse any voltaic cell by the application of an outside source of current and regenerate the active materials. Actually, however, the only such cell in common use is the *storage cell*, used to start an automobile engine. It is both a voltaic cell and an electrolytic cell. While starting the automobile, the storage cell acts as a voltaic cell. When the automobile is running, the storage cell functions as an electrolytic cell.

This cell consists of an electrode of lead and another electrode covered with PbO$_2$. These electrodes are immersed in a sulfuric acid solution, which is the electrolyte.

When the cell is used, that is, when it is acting as a voltaic cell, the lead electrode acts as the anode:

$$Pb \longrightarrow Pb^{++} + 2e^-$$

The lead ion subsequently reacts with the sulfate ion, and insoluble lead sulfate is deposited on the lead plate:

$$Pb^{++} + SO_4^{--} \longrightarrow PbSO_4$$

ELECTROCHEMISTRY

The cathode reaction during the cell's use also produces $PbSO_4$:

$$2e^- + PbO_2 + 4H_3O^+ \longrightarrow 6H_2O + Pb^{++}$$
$$Pb^{++} + SO_4^{--} \longrightarrow PbSO_4$$

When the storage cell operates as an electrolytic cell, that is, when it is being charged, the reverse reactions take place, so that at the lead electrode (now the cathode)

$$PbSO_4 \longrightarrow Pb^{++} + SO_4^{--}$$
$$Pb^{++} + 2e^- \longrightarrow Pb$$

and at the PbO_2 electrode (now the anode)

$$PbSO_4 \longrightarrow Pb^{++} + SO_4^{--}$$
$$Pb^{++} + 6H_2O \longrightarrow PbO_2 + 4H_3O^+ + 2e^-$$

Theoretically the storage cell should last forever, since electrolysis should bring the battery back to its original condition. Unfortunately, slow mechanical disintegration of electrodes limits the life span of the storage cell.

Considerable effort has been made to develop a workable voltaic cell to derive electric energy from gaseous fuels such as hydrogen, carbon monoxide, or hydrocarbons. Known as a *fuel cell*, this type is of great interest at present because it would be a much more efficient method of producing electric energy. In addition, it might possibly eliminate the gasoline engine in automobiles and the consequent air pollution. Finally, this cell is of interest because of its applications to space exploration.

The customary use of fossil fuels (coal and petroleum) as a source of energy is highly wasteful, being no more than 50 percent efficient. Even more serious are the harmful contaminants that these fuels introduce into the atmosphere. The chief pollutant of the atmosphere is carbon monoxide, the deadly nature of which is well known. Yet many deaths occur annually from exhaust fumes which find their way into the interior of a car when the engine is carelessly left running when the car is parked or in the garage.

Other gases which flow out of the automobile exhaust are unburned hydrocarbons, sulfur dioxide, various oxides of nitrogen (principally NO), and lead halides. The latter come from tetraethyllead, which is added as an antiknock agent to ethyl gasoline, and from ethylene dibromide, $Br—CH_2—CH_2—Br$, which prevents the deposition of lead oxides which would be detrimental to the engine. When the exhaust gases of the automobile are trapped in a geographic locality and acted upon by bright sunlight, an extremely complex mixture of materials known as *smog* results. Chemists have labeled ozone, an allotropic form of oxygen, O_3, as the most harmful constituent of smog. Ozone is very reactive, causing bleaching and deterioration of fabrics and rubber and killing vegetation. Trees as far away as 50 miles from Los Angeles are believed to have been killed by smog. The constituents of smog which lead to eye irritation are the peroxyacl nitrates (R—C(=O)—O—O—NO$_2$, where R is a carbon-hydrogen group). Nitrogen dioxide, a brown gas, is responsible for the color of smog and for damage to lung tissue. Other materials found in the mixture include atomic oxygen and free radicals (see Chap. 12).

The fuel cell that has had the most success to date uses hydrogen. This cell consists of inert electrodes (porous carbon) immersed in an alkaline solution. At the anode, hydrogen gas bubbling around the carbon electrode is oxidized:

$$2H_2 + 4OH^- \longrightarrow 4H_2O + 4e^-$$

At the cathode oxygen is reduced:

$$4e^- + O_2 + 2H_2O \longrightarrow 4OH^-$$

The overall reaction of the cell is one of combining hydrogen and oxygen to form water. Recent space flights have used hydrogen-oxygen cells to power the spacecraft as well as to provide a source of water. Some success in developing a fuel cell that can power automobiles has already been achieved. A 27-horsepower truck has been built which is powered by four fuel cells in which hydrazine H:N:N:H (with H H on top) is burned to free nitrogen and water. Since neither product of this fuel cell would be a harmful contaminant to the atmosphere, fuel cells of this type are possible as a partial answer to the air-pollution problem.

An interesting application of the electrolytic cell is *electroplating*, the process of covering one metal with a film of another. Usually this is done to protect the more reactive metal from corrosion and to give it a more attractive appearance. For example, table cutlery can be plated with silver (Fig. 11-4). The spoon or fork to be plated is used as the cathode and a piece of pure silver is used as the anode. The electrolyte used in this process is an aqueous solution containing silver ions. When an electric current is applied, silver loses electrons at the anode and silver ions pick up electrons at the cathode and become free silver on the surface of the cathode.

Metallurgy Metallurgy is the process of extracting metals from their ores. Except for a few inactive metals such as gold and silver, which are obtained from the earth in the free state, metals are extracted from their ores by reducing

METALLURGY

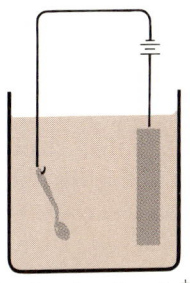

$Ag^+ + e^- \rightarrow Ag \quad Ag \rightarrow Ag^+ + e^-$

Figure 11-4 Electroplating silver on a spoon.

the metallic ions present in naturally occurring compounds to free metals.

Metal ores are most frequently oxides or sulfides but may be carbonates, sulfates, or other compounds. Several physical processes, such as grinding and crushing, are involved in preliminary treatment of the ore. Nonoxide ores are often converted to oxides. The free metal is extracted from the oxide by being heated to a high temperature with a substance which will donate electrons to metallic ions (reducing agent). A common reducing agent is free carbon. For example, in the reaction $Fe_2O_3 + 3C \longrightarrow 2Fe + 3CO$, carbon donates electrons to iron, iron is reduced, and carbon is oxidized.

Often the metal produced by reduction with carbon is insufficiently pure to be used without further treatment. That is, it must be refined. For example, the most important use of copper is as an electrical conductor. However, the presence of even a small amount of impurity greatly detracts from its conducting ability. Metal refining is most often accomplished by electrolytic processes.

Pure copper is obtained in this way. For copper refining the electrolytic cell consists of a cathode of pure copper, an anode of the copper to be purified, and an electrolyte of acidified copper sulfate and sodium chloride solutions. As the electric current is applied, Cu^{++} ions in the solution migrate to the cathode, where they pick up electrons, are reduced to free copper, and begin the buildup of the cathode in layers of pure copper. For the reaction to continue, of course, an equivalent number of Cu^{++} ions must leave the anode and enter the solution. The anode consists of impure copper. Impurities either dissolve into the solution and remain there or fall to the bottom as a sludge. Electrode reactions are:

Anode $\quad Cu \longrightarrow Cu^{++} + 2e^- \quad$ dissolving impure copper
Cathode $\quad Cu^{++} + 2e^- \longrightarrow Cu \quad$ depositing pure copper

Some metals are reduced and purified at the same time by electrolysis. Magnesium metal is produced from molten magnesium chloride:

Anode $\quad 2Cl^- \longrightarrow Cl_2 + 2e^-$
Cathode $\quad Mg^{++} + 2e^- \longrightarrow Mg$

ELECTRON EXCHANGE: OXIDATION AND REDUCTION

Magnesium, the lightest of all structural metals, is alloyed with other metals and used in the manufacture of airplane parts, luggage, ladders, and other articles which must be both light and strong.

Titanium is an unfamiliar metal. This lack of familiarity is not due to rarity (it is many times more abundant than some familiar metals such as copper) but to the difficulty of extracting the metal from its ore. Although titanium is widely distributed in rocks, soil, and other silicate materials, it is extracted commercially only from its oxide, TiO_2. This compound is widely distributed; however, it cannot be reduced with the usual metallurgical reducing agents such as carbon.

The problem of extracting titanium from its ore has been at least partially solved. Twenty years ago this metal sold for about $3,000 per pound. By 1969 its price had been reduced to less than $2 per pound. In general, the extraction is accomplished by converting the oxide to the chloride and reducing the chloride with magnesium. The equations for this extraction are

$$TiO_2 + C + 2Cl_2 \longrightarrow TiCl_4 + CO_2$$
$$TiCl_4 + 2Mg \longrightarrow Ti + 2MgCl_2$$

These reactions take place at such high temperatures that titanium would react with nitrogen and oxygen. Hence this process must be conducted in an atmosphere of argon or some other rare gas.

Titanium is similar in strength to steel; it is less than twice as heavy but more than twice as strong as aluminum, and it is resistant to corrosion. It is already used in the manufacture of airplanes and missiles. More than likely it will be used in the near future for such things as window frames, kitchen utensils, and fishing rods. Clearly it is the metal of the future.

Rate of Reaction

If metal crystals and ions are mixed so that they come in contact with one another, they will react as we have previously described. However, the speed at which these reactions proceed varies from a rate which is imperceptible to one which is explosive. Factors influencing this rate include the nature of the reactants. For example, the reaction between sodium and hydrochloric acid proceeds at a hazardous rate, while lead reacts with hydrochloric acid very slowly. We cannot control the nature of the reactants; however, there are external factors affecting the speed of a reaction that we can control. These include the state of subdivision, concentration, temperature, and the presence of a catalyst. The first two factors may conveniently be discussed here. Discussion of the last two is postponed for the next chapter.

In a heterogeneous reaction, that is, one occurring between reactants in different physical states, the state of subdivision is very important. For example, in the reaction of zinc with 1.0 M hydrochloric acid, the reaction takes place more rapidly if the zinc is in shreds rather than one large chunk because the finely divided zinc crystals expose a larger area to the acid solution and hence more H_3O^+ ions can receive electrons from the zinc crystal at the same time.

We can vividly illustrate the effect of concentration on the speed of the reaction by placing zinc crystals in solutions which have varying concentrations of hydronium ion. Let us consider what happens when an excess of zinc is placed in four different vessels containing equal volumes of (a) 1 M hydrochloric acid, (b) 1 M acetic acid, (c) 0.1 M hydrochloric acid, and (d) 0.1 M acetic acid, respectively (Table 11-2). The ranking of rates at which hydrogen would be evolved from these vessels would be a first, c second, b third, and d last. Hydrochloric acid is 100 percent ionized. Hence the hydronium-ion concentrations in containers a and c would be 1 and 0.1 M, respectively. Acetic acid is less than 2 percent ionized. Thus, the hydronium-ion concentration in containers b and d would be less than 0.02 and 0.002 M, respectively. Therefore the arrangement of the four containers in the order of hydronium-ion concentration would be a first, c second, b third, and d last, corresponding exactly to the arrangement in order of the rate at which hydrogen is evolved. A direct relationship between concentration and rate of reaction is clearly indicated.

TABLE 11-2 Comparison of Four Acid Solutions in the Production of Hydrogen Using an Excess of Zinc and Equal Volumes of Acid

Vessel	Acid	Concentration, M	Percent Ionization	Hydrogen-ion Concentration, M	Rank of Hydrogen-ion Concentration	Rate Rank	Quantity of H_2 Produced
a	HCl	1	100	1	1	1	Same as b; 10 times as much as c and d
b	$HC_2H_3O_2$	1	Less than 2	Less than 0.02	3	3	Same as a; 10 times as much as c and d
c	HCl	0.1	100	0.1	2	2	Same as d; one-tenth as much as a and b
d	$HC_2H_3O_2$	0.1	Less than 2	Less than 0.002	4	4	Same as c; one-tenth as much as a and b

210 ELECTRON EXCHANGE: OXIDATION AND REDUCTION

It should be noted that the same quantity of hydrogen can be obtained from vessels *a* and *b*. However, it will take longer to get this quantity of hydrogen from vessel *b* than from vessel *a*. Similarly, the quantities of hydrogen obtained from vessels *c* and *d* would be the same, but the time required to produce this quantity of hydrogen from *d* would be considerably longer than from vessel *c*. This difference in time requirement is due to the difference in the degree of ionization of the two acids. Although the degree of ionization for acetic acid is much smaller than that for hydrochloric acid, the ionization of the acid continues as hydrogen is produced. That is, the equilibrium $HC_2H_3O_2 + H_2O \rightleftharpoons H_3O^+ + C_2H_3O_2^-$ is shifted to the right as the hydronium ion is removed by the reaction $2H_3O^+ + Zn \longrightarrow Zn^{++} + H_2 + 2H_2O$. Hydrogen is produced as long as acetic acid is present to continue ionizing to form H_3O^+ ions.

The quantity of hydrogen obtained by the reaction of zinc with the acid in vessels *a* and *b* would be 10 times as great as the quantity of hydrogen obtained from the reaction of zinc with the acid in vessels *c* and *d*. Thus we see that although the speed of the reaction is directly related to the hydronium-ion concentration, the quantity of hydrogen produced is not determined by the hydronium-ion concentration but by the total concentration of the acid in the solution.

Questions

1 Name four metals that displace copper ions from solution.
2 Write equations for the following. If no reaction occurs, indicate by writing N.R.
   a $Al + H_3O^+ \longrightarrow$ b $Hg + Au^{3+} \longrightarrow$
   c $Zn + Sn^{++} \longrightarrow$ d $Fe + Zn^{++} \longrightarrow$
   e $Ni + Ag^+ \longrightarrow$ f $Ag + H_3O^+ \longrightarrow$
   g $Co + Cu^{++} \longrightarrow$ h $K + H_3O^+ \longrightarrow$
   i $Cu + Ag^+ \longrightarrow$ j $Sn + Co^{++} \longrightarrow$
3 Design a voltaic cell using iron and silver as electrodes.
   a Which electrode is the anode?
   b At which electrode does reduction take place?
   c What is the direction of the flow of positively charged ions? Negatively charged ions?
   d What is the direction of the flow of electrons?
   e In order to have the maximum voltage, which solution should be most concentrated?
   f Write the equation for the overall cell reaction.
   g What will be the change in mass of the electrodes (if any)?
4 Design an electrolytic cell for the electrolysis of fused KBr.
   a Write equations for the half-reactions at each electrode.
   b At which electrode is oxidation taking place?

c Which electrode is the cathode?
d Write the equation for the overall cell reaction.
e What is the direction of electron flow?

5 What is the fundamental difference between a voltaic cell and an electrolytic cell?

6 In the voltaic zinc-copper cell, what species is oxidized? In the electrolytic cell in which fused NaCl is the electrolyte, what species is oxidized?

7 Why are salt crystals not used as an electrolyte in a cell even though they are completely ionic?

8 For each of the following, indicate which element is oxidized and which is reduced:
a $Mg + 2H_3O^+ \longrightarrow Mg^{++} + H_2 + 2H_2O$
b $Zn + Cu^{++} \longrightarrow Cu + Zn^{++}$
c $Mg + Sn^{++} \longrightarrow Mg^{++} + Sn$
d $CuO + H_2 \longrightarrow Cu + H_2O$
e $Fe_2O_3 + 3C \longrightarrow 2Fe + 3CO$

9 What gases are formed in the dry cell? How are they disposed of?

10 When the storage cell operates as a voltaic cell,
a What is reduced?
b What is oxidized?
c What is the anode?
d What is the cathode?
e What happens to the electrolyte?

11 When the storage cell operates as an electrolytic cell:
a What is reduced?
b What is oxidized?
c What is the anode?
d What is the cathode?
e What happens to the electrolyte?

12 What is a fuel cell?

13 How is smog formed?

14 What are some advantages of the fuel cell over the conventional method of producing electricity?

15 Explain or account for the following:
a One should not attempt to store a solution of $AgNO_3$ in a copper container.
b When an iron nail is placed in a solution of copper sulfate, the nail changes to a metallic copper color.
c Either a solution of NaCl or HCl will perform well as a conductor of electricity. Liquid NaCl (NaCl raised to a sufficiently high temperature so that it melts) is also a good conductor. However, liquid HCl (HCl cooled sufficiently under pressure so that it becomes a liquid) does not conduct a current.

d Some metals which are both abundant and widely distributed are less well known than much less abundant elements.

16 Indicate which of the factors influencing the speed of a reaction are involved in the following:

  a Zinc reacts faster than tin to produce hydrogen from a sulfuric acid solution.

  b Hydrogen is produced faster with zinc and hydrochloric acid than with zinc and acetic acid.

  c Zinc dust reacts faster with hydrochloric acid than zinc sheets.

CHAPTER 12 BREAKING AND FORMING COVALENT BONDS

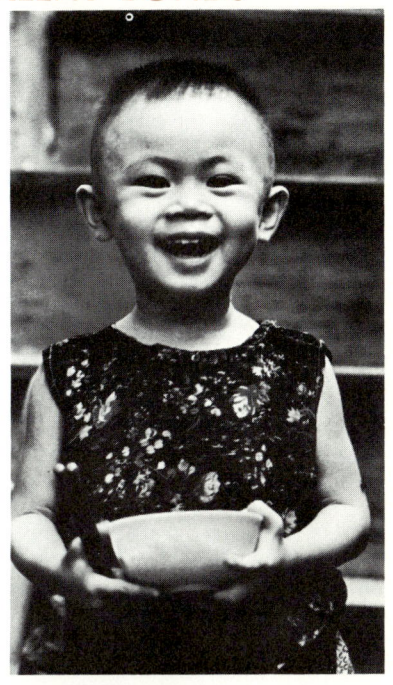

The boy with his empty cereal bowl, who indicates by his expression that he expects it to be filled shortly, illustrates in a dramatic way a major problem—the feeding of the world's growing population. An important aspect of this problem is the production of fertilizers. However, in our concern for increased food production, we must not overlook the problem illustrated by the second picture—the excessive growth of noxious aquatic plants caused by drainage from fertilized lands. (Planned Parenthood and World-Population Foundation and the Federal Water Quality Administration.)

CHAPTER 12 BREAKING AND FORMING COVALENT BONDS

Hydrogen iodide molecules react with each other to form hydrogen and iodine molecules:

$$H\!:\!\ddot{\underset{..}{I}}\!: + H\!:\!\ddot{\underset{..}{I}}\!: \longrightarrow H\!:\!H + \!:\!\ddot{\underset{..}{I}}\!:\!\ddot{\underset{..}{I}}\!:$$

Unlike the reactions we have discussed up to this point, this reaction involves breaking and forming covalent bonds. Which happens first? Do the hydrogen-iodine bonds break before the hydrogen-hydrogen and iodine-iodine bonds are formed? Or do the hydrogen-hydrogen bonds and iodine-iodine bonds form before the hydrogen-iodine bonds are broken? What we are concerned with here is the path or steps by which the reaction proceeds. This is referred to in chemistry as the *mechanism* of the reaction. Determining the mechanism of even very simple reactions is difficult. Despite much activity by chemists in this field, a great deal of work remains to be done. A mechanism is rarely proved. A mechanism, like other theories, is considered correct as long as it agrees with experimental data.

Experimental work supports the following mechanism for the reaction between hydrogen iodide molecules:

1 Two hydrogen iodide molecules collide with sufficient kinetic energy and proper orientation. The molecules must strike laterally so that the two hydrogen atoms are adjacent (see Figs. 12-1 and 12-2).
2 Bonds are formed, bringing into existence the new molecule H_2I_2. Unstable molecules, such as H_2I_2, which are formed as intermediates in a chemical reaction, are called *activated complexes*. The energy required to form this complex is called the *activational energy*. Much of the kinetic energy of the colliding hydrogen iodide molecules is transformed into potential energy in the activated complex.
3 The activated complex can revert to hydrogen iodide, or it can break up into a hydrogen molecule and two iodine atoms. The two iodine atoms quickly combine to form an iodine molecule. Both steps release energy, and the potential energy of the system declines. However, the potential-energy decrease does not bring the system down to the level of the potential energy of hydrogen iodide; hence the reaction is endothermic (see Fig. 12-3).

The reaction we have been discussing is reversible. Hydrogen and iodine combine to form hydrogen iodide. The mechanism here is the reverse of the reaction between hydrogen iodide molecules.

1 Iodine molecules break up into iodine atoms.
2 Two iodine atoms combine with a hydrogen molecule to form H_2I_2.
3 The activated complex forms hydrogen iodide.

216 BREAKING AND FORMING COVALENT BONDS

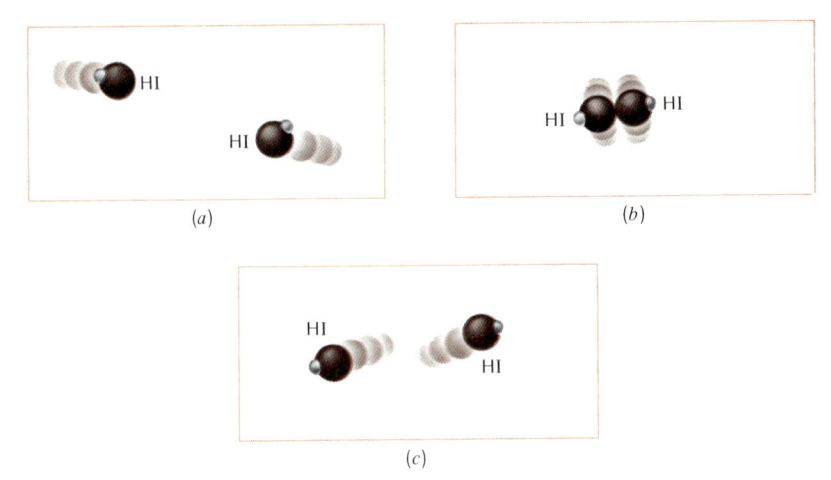

Figure 12-1 An ineffective collision of HI molecules.

In the reaction between hydrogen and chlorine, a very different reaction mechanism prevails. The reaction between hydrogen and iodine is slow and sluggish, but the reaction between hydrogen and chlorine is very rapid—almost explosive. The slowness of the hydrogen and iodine reaction indicates a mechanism of two iodine atoms colliding with one hydrogen molecule.

The hydrogen-chlorine reaction is initiated by exposing a mixture of the two gases to a bright light. Absorption of light of suitable wavelength breaks the chlorine-chlorine bond, producing free atoms. The production

Figure 12-2 An effective collision of HI molecules.

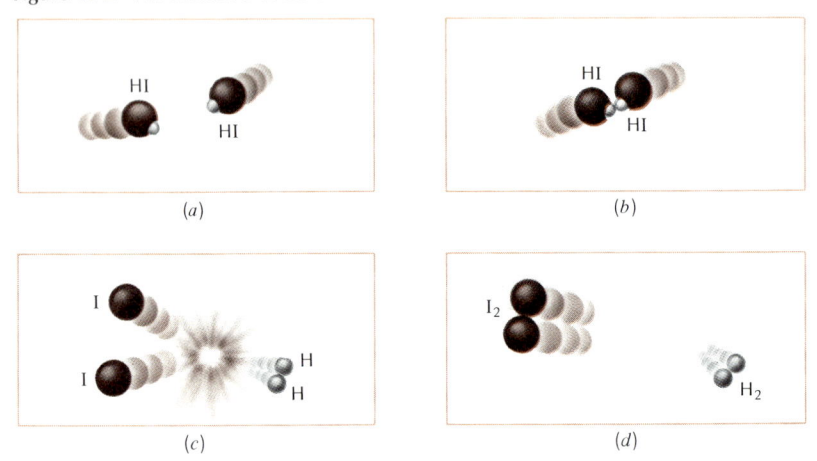

217 BREAKING AND FORMING COVALENT BONDS

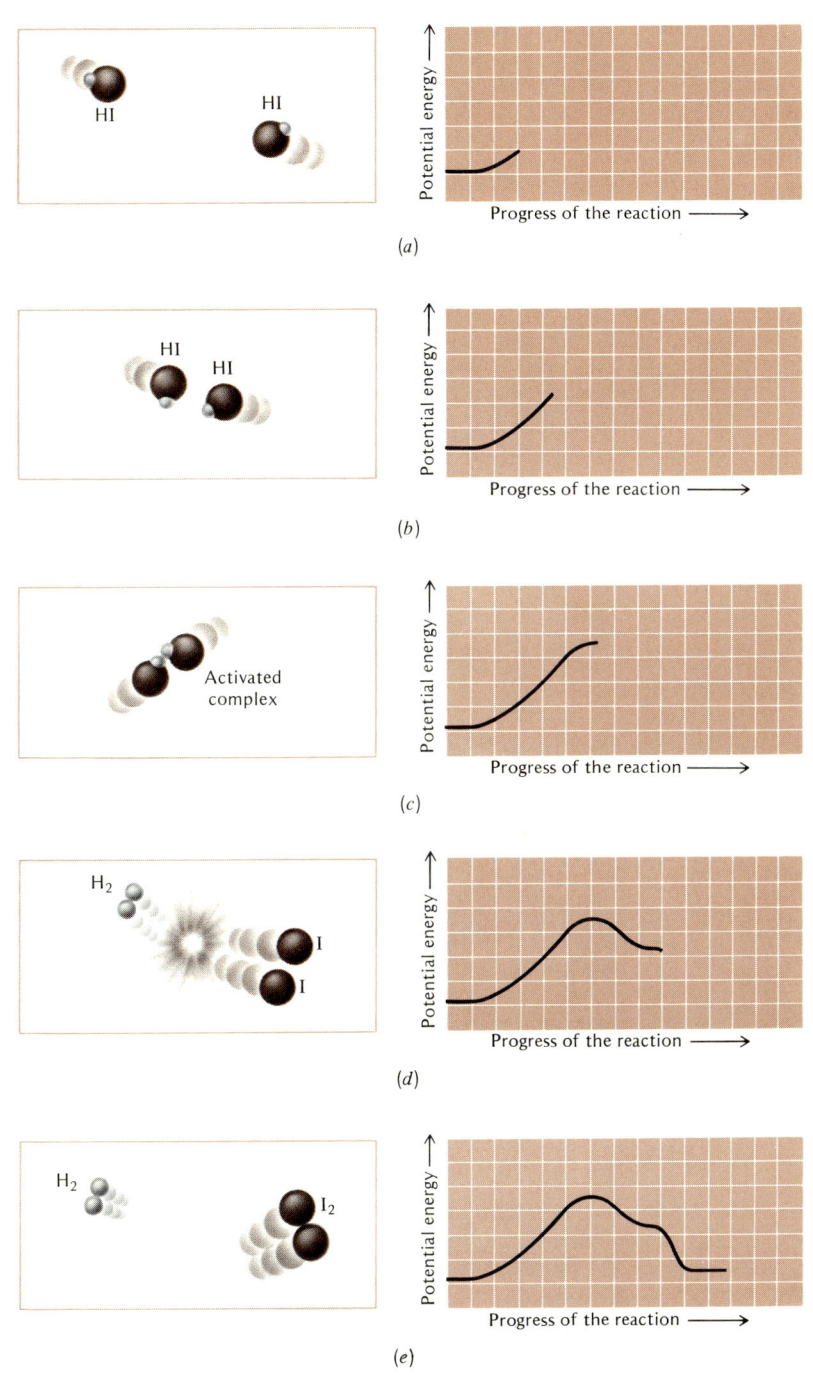

Figure 12-3 Relation of potential energy and reaction progress between two hydrogen iodide molecules.

of the chlorine atoms begins a series of reactions which include the following:

$$:\ddot{\underset{..}{Cl}}:\ddot{\underset{..}{Cl}}: \longrightarrow 2:\ddot{\underset{..}{Cl}}\cdot \tag{1}$$

$$:\ddot{\underset{..}{Cl}}\cdot + H:H \longrightarrow H:\ddot{\underset{..}{Cl}}: + H\cdot \tag{2}$$

$$H\cdot + :\ddot{\underset{..}{Cl}}:\ddot{\underset{..}{Cl}}: \longrightarrow H:\ddot{\underset{..}{Cl}}: + :\ddot{\underset{..}{Cl}}\cdot \tag{3}$$

$$H\cdot + H\cdot \longrightarrow H:H \tag{4}$$

$$:\ddot{\underset{..}{Cl}}\cdot + :\ddot{\underset{..}{Cl}}\cdot \longrightarrow :\ddot{\underset{..}{Cl}}:\ddot{\underset{..}{Cl}}: \tag{5}$$

$$H\cdot + \cdot\ddot{\underset{..}{Cl}}: \longrightarrow H:\ddot{\underset{..}{Cl}}: \tag{6}$$

Reaction (1) initiates the series. Reactions (2) and (3) are propagating reactions. Reactions (4) to (6) are terminating reactions. Since there are a great many more molecules than free atoms in the reaction mixture during the major part of the reaction, reactions (4) to (6) are negligible until the reaction is nearly finished. For most of the reaction time, reactions (2) and (3) alternate. A series of reactions of this type, where the reaction is initiated by some outside influence and then propagated by one or two steps repeated over and over, is called a *chain reaction*.

FREE RADICALS

Free radicals are defined as atoms, ions, molecules, or molecular fragments which contain an unpaired electron. In this category we have a few reasonably stable species, such as NO, NO_2, and ClO_2. Since these species contain an odd number of electrons, they must of necessity contain an unpaired electron. Because they contain an unpaired electron, they are paramagnetic. These few reasonably stable species containing an odd number of electrons are sometimes referred to as *odd molecules*.

Reactions which occur by breaking and making bonds are referred to as *free-radical reactions*. The reaction between hydrogen and iodine as well as the reaction between hydrogen and chlorine are free-radical reactions, and the intermediates in these reactions ($:\ddot{\underset{..}{Cl}}\cdot$, $:\ddot{\underset{..}{I}}\cdot$, and $H\cdot$) are free radicals.

Free radicals are the subjects of intensive investigations all over the world. They have been identified in flames, in the atmosphere, and in industrial reaction containers. Free radicals have also been identified in cigarette smoke and are suspected of causing cancer. The irritation of smog is attributed in part to free radicals.

Most free radicals are much more reactive and less stable than the odd molecules we mentioned previously. Species such as $:\ddot{\underset{..}{Cl}}\cdot$, $H\cdot$, and $:\ddot{\underset{..}{I}}\cdot$ exist only for an instant. *Cryochemistry*, the study of chemical reactions at very low temperatures, is an effective approach to studying free radicals. Extremely rapid cooling of chemical systems to within a few degrees

RATE OF REACTION

above absolute zero prevents the reaction of free radicals and permits their identification. Species containing two unpaired electrons are known as diradicals. The diradical trimethylenemethane, has recently been identified. Its structure has been determined as

Trimethylenemethane

It is reasonably stable at temperatures below 75°K.

Rate of Reaction

In Chap. 11 we discussed the nature of reactants in relation to the speed of a reaction between metal crystals and hydronium ions. We learned that the speed of these reactions is related to the ease with which the metal crystals release electrons. In reactions between molecules, the nature of the reacting materials is also a factor. Here we must consider the ease with which bonds are broken or formed and the mechanism of the reaction. We saw a difference in the reaction rates for reactions involving molecules when we compared the reaction between hydrogen and iodine with the reaction between hydrogen and chlorine. This difference is manifested in part by the mechanisms of the reactions. The hydrogen-iodine mechanism requires that two atoms collide with one molecule, but the hydrogen-chlorine reaction requires that only one atom collide with one molecule. Since the first condition is likely to occur less often than the second, the hydrogen-iodine reaction is slower than the hydrogen-chlorine reaction.

The effect of concentration can also be shown by studying the reaction between molecules. In the case of the hydrogen-iodine reaction it has been found that if the concentration (number of molecules or moles per unit of volume) of either reactant is doubled, the rate of the reaction is doubled. Similarly, if the rate of one reactant is tripled, the rate of the reactant is tripled. Or if the concentration of one component is doubled and the other is tripled, the rate of reaction is 6 times as great; that is, the rate of reaction is proportional to the product of the concentrations of the reactants.

Temperature is an extremely important factor in influencing the speed of a reaction. For many reactions between gases, just an increase of 10°C will double the speed of the reaction. Why should such a small increase in temperature bring about such a large increase in the rate of the reaction? Our first tendency is to explain this phenomenon in terms of increasing the average velocity of the gas molecules and hence increasing the number of collisions between the molecules. However, a temperature increase of 10°C increases the velocity of gas molecules less than 5 percent. Thus the increased number of collisions alone does not explain a doubling of the reaction rate.

The answer to our question lies in the fact that only the most energetic molecules react upon colliding. That is, collision between molecules is not sufficient. We have indicated that proper orientation of the molecules upon collision is also necessary. The third factor is that the molecules must have sufficient kinetic energy upon collision to bring about the formation of new bonds and/or distortion or breaking of old bonds.

The influence of temperature upon the rate of a reaction is best understood by considering Fig. 12-4. We have indicated that the *average* velocity of gas molecules is dependent on the temperature. However, molecules of a given gas do *not* all have the same velocity at a given temperature. Figure 12-4 plots the number of gas molecules with certain velocities at a given temperature. The number of molecules with sufficient kinetic energy to react is indicated by the dashed line. We note that only a small portion of the molecules has sufficient energy for an effective collision. The remaining molecules will not react, even upon collision with the proper orientation. If the temperature is increased to $(T + 10)$°C, the average velocity of the molecules is increased only slightly but the number of molecules with sufficient energy to react upon collision with the proper orientation is approximately doubled. Therefore the speed of the reaction is doubled.

Figure 12-4 Velocity distribution of molecules at T°C and at $(T + 10)$°C. $a =$ molecules with sufficient kinetic energy to react without catalyst at T°C. $a + b =$ molecules with sufficient kinetic energy to react without catalyst at $(T + 10)$°C. $a + d =$ molecules with sufficient kinetic energy to react with catalyst at T°C. $a + b + c + d =$ molecules with sufficient kinetic energy to react with catalyst at $(T + 10)$°C.

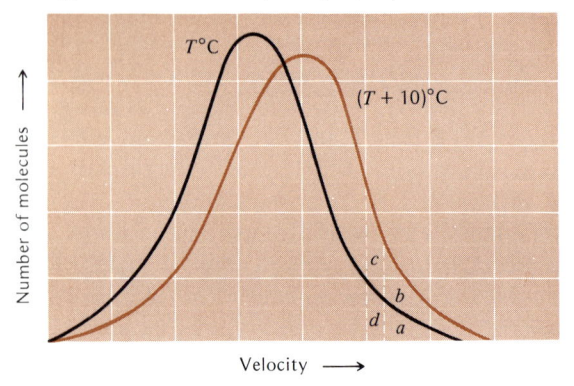

CATALYSTS

The fourth external factor of importance in influencing the speed of the reaction is the presence or absence of a catalyst. A catalyst is a substance which influences the speed of a reaction but can be recovered completely after the reaction is over. Catalysts are often specific for a given reaction. That is, a substance which can act as catalyst for one reaction does not necessarily act as a catalyst for another reaction.

Catalysts do participate in some portion of the mechanism of the reaction. This often involves the formation of an intermediate complex or compound. For example, consider the hypothetic reaction between A and B to form AB. This is a slow reaction. But the reaction between A and C to form AC (A + C $\longrightarrow$ AC) is a very fast reaction. Similarly the reaction between AC and B to form AB and C (AC + B $\longrightarrow$ AB + C) is a rapid reaction. The compound AB is formed more rapidly by the intermediate route involving C than by a direct reaction between A and B. The substance C is completely recovered. Hence C is a catalyst for the reaction between A and B.

The action of a catalyst can be interpreted as one of lowering the activational-energy requirement for the reaction and hence providing a larger portion of molecules of sufficient energy to react upon collision (see Figs. 12-4 and 12-5).

Figure 12-5 A catalyst lowers the activational energy. (1) = activation complex for A + B, (2) = activation complex for A + C, and (3) = activation complex for AC + B.

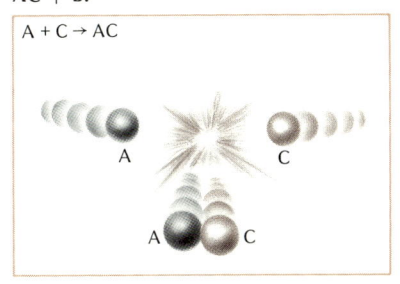

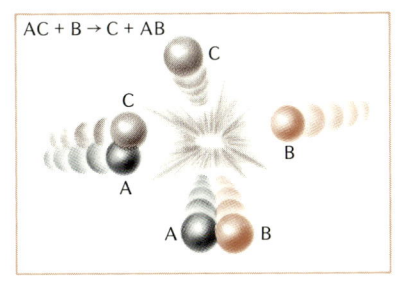

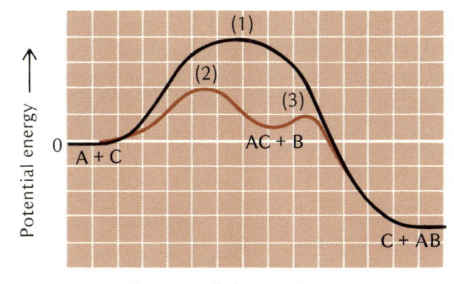

SOME PRACTICAL APPLICATIONS OF CATALYSTS

The use of catalysts to increase the speed of a reaction is of great practical importance. The manufacture of sulfuric acid, the most widely used chemical product, involves a catalyst. The chemistry of the sulfuric acid manufacturing process can be given in three equations:

$$S_8 + 8O_2 \longrightarrow 8SO_2$$
$$2SO_2 + O_2 \rightleftharpoons 2SO_3$$
$$SO_3 + H_2O \longrightarrow H_2SO_4$$

Raw materials for this process are abundantly available. The key to the success of the whole process lies in the second step. Bringing about the reaction of SO_2 and O_2 at a reasonable rate and with economically rewarding yields is the problem. At low temperatures the rate is too slow, but since the reaction is exothermic, increases in temperature cause the equilibrium to shift to the left, thus reducing the yield. An increase in pressure helps, but there are practical limits beyond which pressure cannot be increased. The problem is solved by using catalysts. One catalyst is NO, which combines rapidly with O_2 to form another gas, NO_2:

$$2NO + O_2 \longrightarrow 2NO_2$$

Subsequently NO_2 reacts rapidly with SO_2 to form SO_3 and regenerate NO:

$$NO_2 + SO_2 \longrightarrow SO_3 + NO$$

Thus SO_3 is formed more rapidly by the intermediate route involving NO_2 than by the direct reaction between SO_2 and O_2.

Another important industrial use of a catalyst is in the manufacture of ammonia, a compound essential to agriculture. Although 80 percent of the atmosphere is nitrogen, it is not available to plants, which can utilize only combined or "fixed" nitrogen. The conversion of elemental nitrogen to a combined form is difficult because free nitrogen is extremely unreactive. This inactivity is due to the strength of the triple nitrogen-to-nitrogen bond in the N_2 molecule, :N:::N:. Before the present century man depended on natural sources for combined nitrogen. These sources seemed so scanty and inadequate that some scientists predicted the approaching end of civilization; they saw mankind doomed to starvation by insufficient supplies of nitrogen compounds. The solution to this problem was provided by Fritz Haber, who developed a suitable catalyst for the reaction

$$N_2 + 3H_2 \rightleftharpoons 2NH_3$$

This reaction is exothermic. Thus Haber was confronted with the conflicting requirements of rate and yield. Increased pressure increases both rate and yield, but increased temperature, which increases the rate, shifts the equilibrium to the left with accompanying decrease in yield. As in the manufacture of sulfuric acid, economic success depends upon the use of a suitable catalyst. The catalyst for this reaction has changed a number of times, but it is essentially a mixture of metallic oxides.

The catalyst used in the Haber process is an example of a *contact catalyst*. Solid materials which act as catalysts for reactions between gases hold appreciable quantities of gas on their surfaces. The catalytic effect is probably due in part to the increase in concentration of the reactants. Molecules adsorbed on the surface are crowded closer together than when in the gaseous state. Also in many cases the surface attraction may be sufficiently great to deform or break bonds, so that considerably more reactive species result. This is probably true in the Haber process.

The use of fertilizers such as ammonia has made possible spectacular increases in yield by agriculture, but this remarkable accomplishment has been attended with some unfortunate side effects. Runoff of fertilizers from adjacent agricultural areas has been identified as an important factor in the pollution of inland lakes, such as Lake Erie. Nitrogen and phosphate fertilizers cause aquatic plants to grow prodigiously for a short time. Later, when these plants die, most of the dissolved oxygen in the water is used up when they decompose, bringing death to fish and other forms of marine life.

Our discussion up to this point has been mainly of how the speed of chemical reactions can be increased. However, it is sometimes desirable to slow down chemical reactions. Such is the case in storing food. Food spoilage during storage is often caused by bacteria, but food spoilage can also be due to chemical reactions not involving bacteria. For example, the rancidity of certain fatty foods such as shortenings is caused by the reaction of fats with atmospheric oxygen. We can slow such reactions by altering external factors which influence the speed of chemical reactions. Thus we lower the concentration of atmospheric oxygen by sealing the food in closed containers. We store the food at low temperatures. We make every effort to exclude materials such as copper and iron ions, which accelerate the reaction by catalytic action.

Scientific advancements are not always unmixed blessings. Such is the case with the discovery of the antiknock properties of tetraethyllead, $(C_2H_5)_4Pb$. This discovery is classified as one of the greatest advancements in the improvement of the operation of automobile engines, but tetraethyllead is also responsible for one of the most toxic contaminants in the atmosphere. Gasoline knock is the result of too rapid explosive combustion in the cylinder, giving an uneven jar or knock to the piston head instead of a controlled push. Tetraethyllead slows down the com-

224 BREAKING AND FORMING COVALENT BONDS

bustion of hydrocarbons (gasoline is essentially a mixture of hydrocarbons) through catalytic action. A catalyst which decreases the rate of a chemical reaction is sometimes called an *inhibitor*. It is known that gaseous explosions proceed as chain reactions similar to the reaction between hydrogen and chlorine (see page 218). It is thought that tetraethyllead acts as a chain stopper or inhibitor by eliminating some of the highly reactive species which propagate the reaction.

In addition to producing a highly toxic contaminant of the atmosphere the use of tetraethyllead in gasoline is objectionable in that it prevents the development of a catalytic muffler. This device would recirculate exhaust gases over catalysts consisting of ceramic granules coated with platinum or chromium. With the aid of these catalysts carbon monoxide, hydrocarbons, and nitrogen oxides would undergo reactions producing carbon dioxide, water, and nitrogen. Tetraethyllead interferes with this procedure by coating the catalysts with lead. Substances which destroy the activity of a catalyst are referred to as *catalyst poisons*.

Stoichiometry We have learned that a chemical equation describes a chemical reaction by identifying the reactants and the products. Another important aspect of the chemical equation is that it is a quantitative statement of a chemical reaction. The study of the quantitative aspects of a chemical equation is called *stoichiometry*.

Let us consider the reaction $4HCl + O_2 \longrightarrow 2Cl_2 + 2H_2O$. In addition to telling us that hydrogen chloride and oxygen react to yield chlorine and water, this equation tells us the relative reacting amounts of reactants and products. The coefficients (numbers preceding each reactant and product) indicate the ratio in which the molecules of each reactant react and the molecules of each product are formed. Thus four molecules of HCl react with one molecule of oxygen to yield two molecules of chlorine and two molecules of water. We have learned that there are 6.02×10^{23} molecules in each mole. Thus we also have a quantitative relationship in terms of moles. That is, four moles of HCl react with one mole of oxygen to form two moles of chlorine and two moles of water. In addition to being 6.02×10^{23} molecules, a mole is also 1 gram molecular weight. Therefore, the mole relationship can readily be converted to a gram relationship by multiplying the number of moles by the gram molecular weight. Hence, 146 g of HCl reacts with 32.0 g of oxygen to produce 142 g of chlorine and 36.0 g of water. Finally, if the reactants and products are gases, the equation also expresses a volume relationship. At standard conditions, 1 mole of gas occupies 22.4 liters. Therefore, at standard conditions 4×22.4 liters of HCl reacts with 22.4 liters of oxygen to form 2×22.4 liters of chlorine. Note that any weight and volume units can be used as long as they are the same for each component of the system.

QUESTIONS

The following examples illustrate the quantitative aspects of a chemical equation.

Example 12-1 Consider the equation $2CO + O_2 \longrightarrow 2CO_2$. How many grams of oxygen are needed to react with 112 g of CO?

Solution The gram molecular weight of CO is 12.0 + 16.0, or 28.0 g. Thus 112/28.0 moles of CO are reacting. We see from our equation that 2 moles of CO react with only 1 mole of oxygen. Therefore, 112/28.0 × 0.500 moles of oxygen reacts with 112 g of CO. This number of moles is converted to grams by multiplying by the gram molecular weight of oxygen (16.0 × 2 = 32.0 g). Thus, the number of grams of oxygen reacting with 112 g of CO is 112/28.0 × 0.500 × 32, or 64.0 g.

Example 12-2 How many liters of carbon dioxide measured at standard conditions can be obtained by burning 90.0 g of ethane, C_2H_6? The equation for the reaction is

$$2C_2H_6 + 7O_2 \longrightarrow 4CO_2 + 6H_2O$$

Solution The gram molecular weight of C_2H_6 is 2 × 12.0 + 6 × 1.01, or 30.0 g. Therefore the number of moles of C_2H_6 being burned is 90.0/30.0. From the balanced equation we note that 2 moles of CO_2 is formed for every mole of C_2H_6 burned. Hence 90.0/30.0 × 2 moles of CO_2 is formed upon burning 90.0/30.0 moles of C_2H_6. This number of moles is converted to liters by multiplying by the molar volume at standard conditions, 22.4 liters. Thus, the number of liters of CO_2 formed is 90.0/30.0 × 2 × 22.4 = 134 liters.

Questions
1 What is an activated complex? Give an example.
2 What are four factors that influence the rate of reaction?
3 Indicate which of the factors influencing the speed of a reaction are involved in the following:
   a A bonfire burns faster when it is windy.
   b Hydrogen reacts more rapidly with fluorine than with iodine.
   c Hydrogen peroxide decomposes faster in the presence of MnO_2 than alone at the same temperature.
4 If a gas is evolved by a reaction at 43°C at a rate of 2.0 ml/min, approximately how fast will it be evolved from the same reaction at 73°C?
5 What is an odd molecule? Give several examples.

6 Which of the following are free radicals?
 a Hydrogen molecule
 b Hydrogen ion
 c Hydrogen atom
 d Hydronium ion
 e Chloride ion
 f Hydroxide ion
 g Oxygen atom
 h Oxygen molecule
 i Chlorine atom
 j Chlorine molecule
 k NO_2
 l N_2O_3
 m ClO_2

7 Explain or account for the following:
 a An increase of only 10°C doubles the rate of reaction for some chemical changes.
 b When a camper starts a fire he blows gently on the flame.
 c The reaction between hydrogen and iodine is exothermic, yet energy must be supplied to initiate the reaction.

8 Consider the exothermic reaction for the commercial production of ammonia and explain why:
 a An increase in temperature increases the rate.
 b An increase in temperature decreases the yield.
 c An increase in pressure increases the yield.

9 What information can be obtained from a balanced equation?

10 The equilibrium $N_2 + 3H_2 \rightleftharpoons 2NH_3$ is shifted to the left by an increase in temperature. Yet in the production of ammonia by the Haber process the equilibrium mixture is kept at 500°C. Explain.

11 Give an example of a free-radical reaction.

12 What causes engine knock?

13 What are the objections to using tetraethyllead in gasoline?

14 What is an inhibitor?

15 Consider the reaction $CH_4 + 4Cl_2 \longrightarrow CCl_4 + 4HCl$.
 a How many molecules of chlorine react with 20 molecules of CH_4?
 b How many moles of CCl_4 can be produced from 31.0 moles of methane?
 c How many molecules of CH_4 are needed to produce 44 molecules of HCl?
 d How many grams of CCl_4 can be produced from 30.0 moles of Cl_2?
 e How many molecules of CCl_4 can be produced from 10.0 g of CH_4?
 f How many grams of Cl_2 will react with 93.0 g of CH_4?
 g What volume of HCl measured at standard conditions will be produced from 39.0 g of Cl_2?
 h If 47.0 g of CCl_4 is produced in a reaction, how many grams of HCl are produced at the same time?
 i If 33 molecules of CH_4 and 68 molecules of Cl_2 are mixed and allowed to react, how many molecules of HCl are produced?

CHAPTER 13 TRANSFORMATIONS OF MATTER BY THE LOSS OR GAIN OF NUCLEONS

Scientific instruments placed on the moon on November 19, 1969, by astronauts Charles Conrad and Alan Bean received their power from a compact nuclear generator. (National Aeronautics and Space Administration.)

CHAPTER 13 TRANSFORMATIONS OF MATTER BY THE LOSS OR GAIN OF NUCLEONS

Rutherford's alpha-particle scattering experiments established the theory of the nuclear nature of the atom by demonstrating that most of the mass of the atom is concentrated in its positively charged nucleus. It has been further postulated that the nucleus consists entirely of neutrons and protons, collectively called *nucleons*. That these nucleons must be bound together by forces very different from electrostatic forces (and considerably stronger) becomes immediately apparent. Electrostatic repulsions between the positive protons should cause the nucleus to fly apart. The very fact that the nucleus exists, then, indicates the presence of other, greater forces.

Rutherford and subsequent experimenters determined the relative size of the atomic nucleus. Approximately 10^{-15} of the volume of the atom is occupied by the nucleus. The radius of the nucleus is on the order of 10^{-13} cm. The radius of an atom is of the order of 10^{-8} cm, or 100,000 times the nuclear radius. The distortion intrinsic to many models and diagrams purporting to show the structure of the atom should be apparent. If the period at the end of this sentence were an atomic nucleus, the atom in which it exists would have a diameter exceeding the length of a football field.

In addition to the apparent contradiction of electrostatic theory by a nucleus consisting wholly of positive and neutral particles, another startling fact emerges. If most of the mass of the atom is concentrated in 10^{-15} of its volume, what is the density of nuclear material? Calculation yields densities on the order of hundreds of trillions of grams per cubic centimeter, or billions of tons per cubic inch. At this point we might be inclined to dismiss the nature of the nucleus and its binding forces as so unusual as to be unexplainable in our terms. But there are still several facts which will render it at least more understandable.

The extremely high density of nuclear material is due to the fact that most of the atom is space. If all the atoms of the earth collapsed into their nuclei, the earth would shrink to a sphere whose diameter would be little more than the length of three football fields. In stars called *white dwarfs* this collapse has apparently taken place to some degree, for they have densities many thousands of times greater than those observed on earth.

The fact that at the very short distances of the atomic nucleus there are forces so strong that electrostatic repulsions or attractions may be neglected is a situation parallel to one already encountered. Gravitational forces, the attraction of every object for every other object, and most noticeably the attraction between the earth and all objects, are responsible for our concept of weight. Moreover, these forces give our world its "up" and "down" and are the reason phrases like "keeping one's feet on the ground" indicate the normal state of affairs. However, when considering the electrostatic forces binding the electron to the atom, the student will remember that nowhere in that discussion was reference made to the "top" or the "bottom" of the atom. In fact, when electrostatic forces which are effective over distances on the order of atomic radii (10^{-8} cm) are considered, gravitational forces are completely ignored. The reason for this lies in the relative magnitude of the two forces at various distances. At ordinary distances, measured in feet or inches or even in fractions of a centimeter, gravitational forces are all that need to be considered. Electrostatic forces are relatively short-range forces and almost nonexistent at these distances. When we consider interactions at distances of 10^{-8} cm, however, the roles are reversed. It can be shown by experiment that at these distances the electrostatic attraction between an electron and a proton is 10^{40} times as great as their gravitational attraction for each other.

A similar situation develops when we consider relationships at distances of 10^{-13} cm, in that now the extremely short-range nuclear forces, which were ineffective at the distance of the atomic radius, have a magnitude many times that of the electrostatic forces.

Thus, we have described three forces, each responsible for the relationships between objects at different ranges. Our everyday world with its normal distance relationships is governed by gravitational forces. In the world of the atom, electrostatic forces dominate, but nuclear relationships are determined by the tremendously larger *nuclear forces*.

Nuclear Reactions: Symbols and Equations

Discussion of atomic nuclei and the transformations in which they are involved requires a new use of the chemical symbol. All atoms of a given element have the same atomic number, which corresponds to the number of protons in the nucleus. Atoms of the same element, however, may have different numbers of neutrons in the nucleus, and thus, different mass numbers. Atoms of the same element with different mass numbers are called isotopes. In the lower left-hand corner of the chemical symbol is written the atomic number, and in the upper left-hand corner, the mass number. Symbols for the three common isotopes of oxygen are $^{16}_{8}O$, $^{17}_{8}O$, and $^{18}_{8}O$.

The forces which bind together the nucleus of the atom depend upon the relative numbers of protons and neutrons in the nucleus. The more protons there are in a nucleus, the greater the electrostatic repulsion will be; the greater the total number of nucleons, the greater the attractive forces. As the atomic number increases, more and more neutrons are required for stability. Many of the stable atoms of lower atomic number contain neutrons and protons in a 1 : 1 ratio, as $^{12}_{6}C$, $^{24}_{12}Mg$, $^{28}_{14}Si$, $^{40}_{20}Ca$. As the atomic number increases, the neutron-to-proton ratio also increases, as shown in $^{80}_{35}Br$, $^{127}_{53}I$, $^{137}_{56}Ba$.

NUCLEAR REACTIONS: SYMBOLS AND EQUATIONS

NATURAL RADIOACTIVITY

With over a thousand known atoms, many have nuclei in which the neutron-to-proton ratio is less than optimum for stability. However, of the 325 naturally occurring atoms, 260 are stable and only 65 are unstable. Unstable nuclei disintegrate spontaneously in transformations called *radioactivity*. The scientific observation of these transformations came very late in chemical history (toward the end of the nineteenth century, in fact), for the very good reason that radioactivity is completely undetectable by the five senses. Only when a nuclear transformation has been utilized to bring about a secondary reaction, such as phosphorescence on a zinc sulfide screen or the exposure of a photographic film or the ionization of a gas as in the Geiger counter, are we able to detect its presence.

Using these and other methods of detection, scientists have learned much about the nature of the spontaneous nuclear reactions known as radioactive decay. In general, the decay consists of a small explosion in which one or more particles are ejected from the nucleus. These particles are either an electron, called in this context a *beta particle*, or an alpha particle, already described as consisting of two neutrons and two protons.

Predicting the effect of an alpha-particle emission upon an atom is a matter of simple arithmetic if we make use of chemical symbols to represent the transformation. For the alpha particle we shall use the symbol ^4_2He. If an atom of radium 226 emits an alpha particle, what remains?

$$^{226}_{86}Ra \longrightarrow {^4_2}He + ?$$

The atom which remains will have 84 protons and 138 neutrons, giving it a mass number of 222. The element with atomic number 84 is polonium, Po. The reaction should be written

$$^{226}_{86}Ra \longrightarrow {^4_2}He + {^{222}_{84}}Po$$

Other atoms which emit alpha particles are polonium 210, thorium 230, and uranium 238. Equations representing these transformations are

$$^{210}_{84}Po \longrightarrow {^4_2}He + {^{206}_{82}}Pb$$
$$^{230}_{90}Th \longrightarrow {^4_2}He + {^{226}_{88}}Ra$$
$$^{238}_{92}U \longrightarrow {^4_2}He + {^{234}_{90}}Th$$

The description of the nucleus as consisting of neutrons and protons made no mention of electrons in the nucleus. How then does it account for beta-particle decay? Beta emission converts $^{214}_{82}Pb$ into $^{214}_{83}Bi$ and converts $^{210}_{83}Bi$ into $^{210}_{84}Po$. We might conclude from this that beta emission converts a nuclear neutron into a proton and an electron, the latter being then emitted. It would perhaps be better to say that a beta particle emission *has the effect* of converting a nuclear neutron to a proton, for the nature of the changes within the nucleus is much more complex than this simple statement would imply. The symbol for the beta particle is $_{-1}^{0}e$, indicating negative charge and no appreciable mass. Nuclear equations to represent transformations by beta-particle emission are

$$_{82}^{214}Pb \longrightarrow {_{-1}^{0}e} + {_{83}^{214}Bi}$$
$$_{83}^{210}Bi \longrightarrow {_{-1}^{0}e} + {_{84}^{210}Po}$$

A third form of natural radioactivity is the emission of gamma rays, which are extremely short-wavelength, high-energy radiations. The unstable nuclei which emit an alpha or a beta particle apparently undergo some subsequent rearrangements as they settle into a state of greater stability (lower energy). It is in the course of these rearrangements that energy is emitted in the form of gamma rays.

HALF-LIFE

Natural radioactivity is completely unaffected by variations in conditions which affect other types of transformations. A sample of radium undergoes decay at a rate which is completely independent of temperature, pressure, state of subdivision, or physical state. The rate depends upon one variable only, and that is the amount of radium present. To express the rate of decay of a radioactive isotope, scientists use the concept of the *half-life*. A half-life is the length of time it takes for one-half of any given sample of a substance to decay. The half-life of bismuth 210 is 5 days. This means that given a sample of this isotope weighing 0.200 g, in 5 days the sample will contain only 0.100 g of bismuth 210, and in 10 days only 0.050 g, and in 15 days only 0.025 g, and so on. Half-lives vary from one extreme to the other. The half-life of polonium 212 is 3.0×10^{-7} sec, and the half-life of thorium 232 is 13.9 billion years.

ARTIFICIAL NUCLEAR TRANSFORMATIONS

As soon as man began observing radioactive decay and recognizing it for what it is—the actual transformation of one element into another—he began considering ways of utilizing this new process to discover more about the atom and its nucleus. Could nuclear transformations be induced? Again the name of Rutherford is associated with the major step forward. In 1918, while subjecting various gases to the passage of alpha particles, he observed and verified the following transformation:

$$_{7}^{14}N + {_{2}^{4}He} \longrightarrow {_{1}^{1}H} + {_{8}^{17}O}$$

NUCLEAR REACTIONS: SYMBOLS AND EQUATIONS

He deduced from his observations that the alpha particle had penetrated the nitrogen nucleus, rendering it unstable. The transformation which followed involved the emission of a fast-moving proton, and thus the nucleus remaining was an entirely different element, an isotope of oxygen. An equation to show the intermediate or *compound nucleus* would be

$$^4_2He + ^{14}_7N \longrightarrow (^{18}_9F) \longrightarrow ^{17}_8O + ^1_1H$$

Experiments soon produced similar reactions with other *target nuclei*. Some of them were

$$^{19}_9F + ^4_2He \longrightarrow (^{23}_{11}Na) \longrightarrow ^{22}_{10}Ne + ^1_1H$$
$$^{27}_{13}Al + ^4_2He \longrightarrow (^{31}_{15}P) \longrightarrow ^{30}_{14}Si + ^1_1H$$
$$^{32}_{16}S + ^4_2He \longrightarrow (^{36}_{18}Ar) \longrightarrow ^{35}_{17}Cl + ^1_1H$$

The observation of nuclear transformations was improved with the development of the Wilson cloud chamber, a schematic diagram of which is given in Fig. 13-1. It consists of a cylinder filled with a gas saturated with water vapor. As the piston is moved down, rapid expansion cools the gas and it becomes supersaturated with water vapor. Supersaturation is a metastable state, and if there are any ions in the chamber, water vapor will condense upon them, forming tiny droplets. Thus the tracks of charged particles entering from the right can be seen and photographed—usually from two angles at the same time in order to determine the plane in which the tracks are formed.

Alpha-particle bombardment of the elements boron and beryllium yielded a highly energetic radiation. Believed at first to consist of gamma rays, it was of much higher energy than had previously been observed

Figure 13-1 Schematic diagram of Wilson cloud chamber.

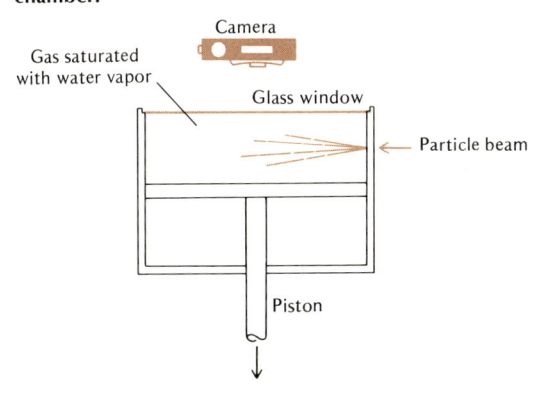

234 TRANSFORMATIONS OF MATTER BY THE LOSS OR GAIN OF NUCLEONS

for them. It possessed great penetrating power, being able to pass through a 1-in. block of lead and lose only one-half of its intensity. When this radiation was directed at a block of paraffin, however, another phenomenon was observed. Protons were ejected from the paraffin. This led James Chadwick to suggest that the radiation from the boron and beryllium consisted not of gamma rays but of streams of particles. Their great penetrating power was due, according to him, to the fact that they carried no charge and thus were able to pass through the collection of positive and negative particles which constitute matter with little interaction. Direct collision with lighter nuclei such as the proton, however, would result in the latter's being ejected. The particle thus described was the neutron, the third fundamental particle. In 1935 Chadwick was awarded the Nobel prize for his discovery.

Subsequently scientists learned that artificial radioactivity might consist of the emission of rays of other particles. Among those identified was the *positron*, a particle of mass equivalent to the electron but carrying a positive charge.

Soon scientists were using many particles to bombard nuclei and then constructing instruments to increase the effectiveness of their nuclear bullets. Linear accelerators, the cyclotron, the betatron, and the synchrotron, were built to give greater kinetic energy to the charged particles used in probing the nucleus. Many data were accumulated, and in the early 1940s, the periodic table was extended as the *transuranic elements* (atomic number 93, 94, . . .) were produced, one by one.

The New Elements Nuclear transformations of this kind have resulted in 16 man-made elements. Those man-made elements heavier than uranium (all but three) are referred to as the transuranic elements. Nuclear transformations resulting in the production of 11 of the 13 transuranic elements are represented by the following nuclear equations:

$$^{238}_{92}U + ^{2}_{1}H \longrightarrow ^{239}_{93}Np + ^{1}_{0}n$$

$$^{238}_{92}U + ^{4}_{2}He \longrightarrow ^{239}_{94}Pu + 3\,^{1}_{0}n$$

$$^{239}_{94}Pu + ^{4}_{2}He \longrightarrow ^{240}_{95}Am + ^{1}_{1}H + 2\,^{1}_{0}n$$

$$^{239}_{94}Pu + ^{4}_{2}He \longrightarrow ^{242}_{96}Cm + ^{1}_{0}n$$

$$^{244}_{96}Cm + ^{4}_{2}He \longrightarrow ^{245}_{97}Bk + ^{1}_{1}H + 2\,^{1}_{0}n$$

$$^{238}_{92}U + ^{12}_{6}C \longrightarrow ^{246}_{98}Cf + 4\,^{1}_{0}n$$

$$^{238}_{92}U + ^{14}_{7}N \longrightarrow ^{247}_{99}Es + 5\,^{1}_{0}n$$

$$^{238}_{92}U + ^{16}_{8}O \longrightarrow ^{249}_{100}Fm + 5\,^{1}_{0}n$$

$$^{253}_{99}Es + ^{4}_{2}He \longrightarrow ^{256}_{101}Md + ^{1}_{0}n$$

$$^{246}_{96}Cm + ^{13}_{6}C \longrightarrow ^{254}_{102}No + 5\,^{1}_{0}n$$

$$^{252}_{98}Cf + ^{10}_{5}B \longrightarrow ^{257}_{103}Lw + 5\,^{1}_{0}n$$

235 THE NEW ELEMENTS

It has been postulated that the transuranic elements existed in nature at the beginning of time but all disappeared from this planet through radioactive decay, except perhaps for traces of plutonium and neptunium. They may exist in stars, where they may be constantly replenished through various nuclear transformations.

Russian scientists announced in 1964 that they had synthesized element 104 by the bombardment of plutonium nuclei with nuclei of neon ($^{242}_{94}Pu + ^{22}_{10}Ne \longrightarrow ^{260}_{104}X + 4\,^{1}_{0}n$) and had named it kurchatovium after the Russian scientist Igor Kurchatov. If the Russian accounts are true, this is the first transuranic element not discovered by an American. So far, American scientists have not been able to duplicate the Russian results in the production of element 104. However, in 1969 scientists at Berkeley, California, announced the synthesis and identification of this element by other means. The nuclear transformation used by the California scientists is represented by the nuclear equation

$$^{249}_{98}Cf + ^{12}_{6}C \longrightarrow ^{257}_{104}X + 4\,^{1}_{0}n$$

Scientists at the Lawrence Radiation Laboratory in Berkeley announced in May 1970 the discovery of element 105. To make element 105, the scientists bombarded a 6 × 10⁻⁵-g target of californium 249 with a beam of nitrogen 15 nuclei in the heavy-ion linear accelerator (HILAC):

$$^{249}_{98}Cf + ^{15}_{7}N \longrightarrow ^{260}_{105}X + 4\,^{1}_{0}n$$

The isotope of element 105 thus produced proved to be more stable, that is, to have a much longer half-life, than expected. This has engendered considerable optimism, so that predictions that elements 106 and 107 may be in existence in the near future are being made.[1]

Scientists are now thinking about element 114. This element, which will be chemically similar to lead, may prove to be much more stable than any of the transuranic elements synthesized so far, for it is expected that a region of nuclear stability will be encountered in the vicinity of atomic number 114. Computations show that stability will decrease again after element 114, until an atomic number in the vicinity of 126 is reached.

As new elements are discovered, the periodic table must be extended to accommodate them. Element 103 is the last element of the second rare-earth series, indicating that with this element the $5f$ subenergy level is filled with 14 electrons. Element 104 begins a new transition series with the "last" electron of each atom going into the $6d$ subenergy level. This

---

[1] American scientists have suggested that element 104 be named rutherfordium and that element 105 be named hahnium, in honor, respectively, of Ernest Rutherford, British Nobel prize recipient for his work with the chemistry of radioactive substances, and Otto Hahn, the German scientist who received the Nobel prize for the discovery of nuclear fission.

TABLE 13-1 Periodic Table Showing Positions of Elements Beyond 103

s BLOCK REPRESENTATIVE

	IA	IIA
1s	1 H	
2s	3 Li	4 Be
3s	11 Na	12 Mg
4s	19 K	20 Ca
5s	37 Rb	38 Sr
6s	55 Cs	56 Ba
7s	87 Fr	88 Ra
8s	119	120

d BLOCK TRANSITION

	IIIB	IVB	VB	VIB	VIIB	VIII			IB	IIB
3d	21 Sc	22 Ti	23 V	24 Cr	25 Mn	26 Fe	27 Co	28 Ni	29 Cu	30 Zn
4d	39 Y	40 Zr	41 Nb	42 Mo	43 Tc	44 Ru	45 Rh	46 Pd	47 Ag	48 Cd
5d	57 La	72 Hf	73 Ta	74 W	75 Re	76 Os	77 Ir	78 Pt	79 Au	80 Hg
6d	89 Ac	104	105	106	107	108	109	110	111	112
7d	121									

p BLOCK REPRESENTATIVE

	IIIA	IVA	VA	VIA	VIIA	0
1s						2 He
2p	5 B	6 C	7 N	8 O	9 F	10 Ne
3p	13 Al	14 Si	15 P	16 S	17 Cl	18 Ar
4p	31 Ga	32 Ge	33 As	34 Se	35 Br	36 Kr
5p	49 In	50 Sn	51 Sb	52 Te	53 I	54 Xe
6p	81 Tl	82 Pb	83 Bi	84 Po	85 At	86 Rn
7p	113	114	115	116	117	118

f BLOCK RARE EARTHS

4f	58 Ce	59 Pr	60 Nd	61 Pm	62 Sm	63 Eu	64 Gd	65 Tb	66 Dy	67 Ho	68 Er	69 Tm	70 Yb	71 Lu
5f	90 Th	91 Pa	92 U	93 Np	94 Pu	95 Am	96 Cm	97 Bk	98 Cf	99 Es	100 Fm	101 Md	102 No	103 Lw
6f	122	123	124	125	126									

236

FISSION AND FUSION REACTIONS

transition series will end with element 112. Elements 113 through 118 will be representative elements (see Table 13-1).

Fission and Fusion Reactions

Investigations employing neutrons as projectiles led scientists into another area, that of atomic fission. It was discovered that when certain nuclei are struck by a neutron, a splitting apart, or fission, occurs.

Uranium 235, when bombarded by neutrons, breaks up into several smaller nuclei. Some of the possible fission products are shown in these equations:

$$^{235}_{92}U + ^{1}_{0}n \longrightarrow ^{90}_{36}Kr + ^{144}_{56}Ba + 2\,^{1}_{0}n$$
$$^{235}_{92}U + ^{1}_{0}n \longrightarrow ^{93}_{35}Br + ^{140}_{57}La + 3\,^{1}_{0}n$$

There are two all-important characteristics of a fission reaction. The first is the concurrent emission of neutrons. Since it was a neutron which initiated the reaction, the ability to be self-sustaining is evidenced. This type of reaction is called a chain reaction. Neutrons are relatively short-lived, however, and unless fissionable nuclei are present in sufficient quantity and within range, the reaction will quickly cease. The term *critical mass* refers to that minimum amount of fissionable material which must be present in order for a reaction to be self-sustaining. In nuclear reactors, amounts of fissionable material in excess of the critical mass are present, but neutron-absorbing material in the form of control rods is introduced into the material in order to make it possible to control the rate of the reaction.

The second characteristic of the fission reaction is the release of large quantities of energy. Amounts of energy released in fission reactions are millions of times greater than the amounts released by chemical reactions involving similar amounts of material. What is the source of these tremendous amounts of energy?

To explain, we must first consider the mass relationships involved in some simpler nuclei. The lithium 7 atom consists, according to atomic theory, of three protons, four neutrons, and three electrons and has an atomic weight of 7.01601. However, if we sum the masses of these 10 constituent particles, and multiply by Avogadro's number, we obtain 7.05813 g/mole as the predicted gram atomic weight:

$3(9.1091 \times 10^{-28}$ g/electron$) + 3(1.67252 \times 10^{-24}$ g/proton$)$
$\qquad + 4(1.67482 \times 10^{-24}$ g/neutron$) = 1.171957 \times 10^{-23}$ g/atom
$(1.171957 \times 10^{-23}$ g/atom$)(6.02252 \times 10^{23}$ atoms/mole$) = 7.05813$ g/mole

This figure is $7.05813 - 7.01601$, or 0.04212 g more than the actual atomic weight. This mass has been lost when the atom was produced.

It was lost by being converted into energy according to Einstein's relation

$$E = mc^2$$

This is called the *binding energy* of the nucleus, and may be used to give some idea of the stability of nuclei relative to each other. The binding energy of the nucleus is the energy which would have to be supplied in order to break the atom apart into its separate particles. A plot of the relationship between binding energy and mass number is given in Fig. 13-2. Note that the relative binding energy per nucleon increases with increasing mass number until the region around mass numbers of 60 to 70 is reached, after which it begins to decrease. Correctly interpreted, this would tell us that an atom of carbon 12 has a higher binding energy than an atom of lithium 7 and thus would require a larger amount of energy for disintegration. It would also imply that an atom of iron 56 is more stable than any of the lighter metals. But what do the decreasing values for binding energies in the right-hand portion of the graph signify? These tell us that an atom of bromine 93 or one of krypton 90 is more stable than an atom of uranium 235, with an evolution of energy accompanying the splitting of a uranium 235 atom into bromine 93 and a krypton 90 atom. This is the source of the energy of fission reactions. The products are more stable than the original reactant, and so the reaction proceeds with evolution of energy.

The binding-energy curve of Fig. 13-2 indicates that not only reactions by which we move from one position toward the center on the right-hand side of the curve will evolve energy but also reactions by which we move

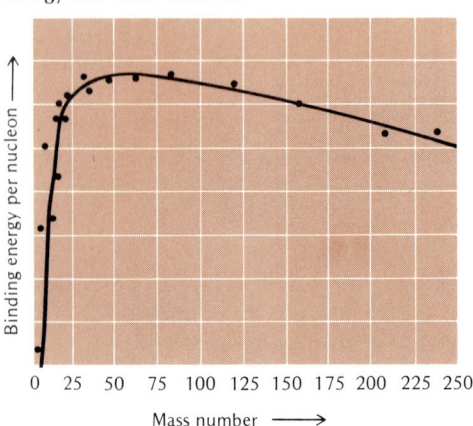

Figure 13-2 The relationship between binding energy and mass number.

from the left side toward the center. This type of reaction is exemplified by combining two light nuclei into one heavier one, as $_1^2H + {}_1^3H \longrightarrow {}_2^4He + {}_0^1n$.

This is the equation for the reaction of the *hydrogen bomb*, in which a deuteron, $_1^2H$, and a triton, $_1^3H$, the positive ions of the heavy isotopes of hydrogen, are fused into a helium nucleus and a neutron. Activational energy requirements for fusion reactions necessitate temperatures as high as 10^8°C. The only known method at present to attain this intensity of heat is in a fission reaction, so that a fission-type reaction is used to trigger or initiate the fusion reaction.

Elements on the left-hand portion of the binding-energy curve are farther apart in energy than those on the right-hand portion. Thus it is that fusion reactions are sources of even greater amounts of energy than fission reactions. The present theory to account for the energy of the sun suggests that a conversion of mass to energy through a fusion type reaction of hydrogen to helium is responsible.

Awesome amounts of energy through nuclear reactions have become available to man at a time when he is being forced to consider the possible exhaustion of the traditional energy sources, the fossil fuels, coal, oil, and gas. Atomic energy may solve that problem, but as with any technological advance, it creates new ones. The testing of fission reactions has produced much controversy. The debris of an atomic explosion moves into the atmosphere and falls out slowly over the entire globe. Most fission products have relatively short half-lives, so that dangerous radioactivity is not of long duration. But some fission products, notably strontium 90, have very long half-lives. Not only does strontium 90 have a half-life of 25 years, but because it is chemically similar to calcium (refer to its periodic table position) it is incorporated into plants, then into human food, and finally into the body, where it lodges chiefly in the bones. Research has revealed a great increase in the amount of strontium 90 in human beings over the past 15 years. To keep this possible danger in its proper perspective, scientists also tell us that fallout radiation absorbed by the body is only approximately 10 percent of that absorbed from natural sources and that in a year the amount of fallout radiation absorbed is about equal to 10 percent of that received in a chest x-ray. Although at present a relatively minor danger, it is one which scientists feel the need to monitor carefully.

Nuclear power plants may help to alleviate another problem, the pollution of the atmosphere by the combustion products of the traditional fuels. It creates other pollution problems, however. The disposal of radioactive wastes requires some ingenious methods. Millions of gallons of radioactive liquids have been sealed in cement and dropped to the ocean floor. Atomic plants also have a thermal pollution problem, because the tremendous quantities of water used as coolant, if discharged back into the river or lake from which they were taken, will by raising the temperature threaten the marine life of the area.

Problems like these will be solved in the course of time. Atomic energy already powers submarines, supplies electric power to whole communities, and produces fresh water from the sea. The possibilities are limited only by man's ability to control and direct this energy toward the betterment of mankind rather than to his destruction.

Questions

1 What are some ways of detecting the presence of radioactive materials?
2 How would a positron emission affect the atomic number? How would it affect the mass number?
3 How can a radioactive element disintegrate to an element with a lower atomic number? A higher atomic number?
4 Why do neutrons penetrate the nucleus more readily than protons?
5 What is the name of a particle emitted by a man-made radioactive element that is not emitted by naturally occurring elements?
6 What do the reactions of hydrogen with chlorine to form hydrogen chloride and the fission of uranium 235 nuclei have in common?
7 Explain the operation of the Wilson cloud chamber.
8 Prepare a chart of the nuclear-emission particles described in this chapter. For each one, give the name, symbol, charge, and mass (if given).
9 How does the sun get its energy?
10 How would you explain the following?
  a It is impossible to have a pure sample of uranium or any other radioactive element.
  b Beta particles are emitted from the nucleus which has been described as consisting of neutrons and protons only.
  c Both nuclear fusion and fission can bring about the release of energy.
11 Write a balanced nuclear equation for alpha-particle emission from the following:
  a Radium 226   b Uranium 234
  c Uranium 238   d Actinium 227
  e Thorium 230   f Polonium 218
12 Write a balanced nuclear equation for beta-particle emission from the following:
  a Carbon 14   b Sulfur 36
  c Thorium 234   d Lead 214
  e Lead 210   f Bismuth 210

13 Polonium 210 has a half-life of 138 days. How long would it take a 1.00-g sample of polonium to decay to 0.25 g?
14 If 2.0 g of an unknown isotope decays to 0.25 g in 15 hr, what is its half-life?
15 Write balanced nuclear equations to illustrate the following:
   a Magnesium 23 emits a positron.
   b Neon 19 emits a positron.
   c Copper 24 emits a positron.
   d Aluminum 28 emits a beta particle.
   e Polonium 212 emits an alpha particle.
16 Supply the missing particle or particles:
   a Oxygen 18 captures a proton and forms a neutron and _____.
   b Rubidium 85 captures a neutron and forms a proton and _____.
   c Fluorine 19 captures a neutron and forms nitrogen 16 and _____.
   d Aluminum 27 captures an alpha particle and forms a proton and _____.
   e Nitrogen 14 captures a neutron and forms nitrogen 13 and _____.
17 In each case below refer to Fig. 13-2 and predict which of the two nuclei should be the more stable:
   a B or Mn   b I or Br
   c Hg or Be   d Mg or Bi
18 Three natural radioactive decay series are known. One of these starts with uranium 238 and culminates in the stable isotope lead 206. Write formulas for the 13 intermediate products if the order of particle emission is alpha, beta, beta, alpha, alpha, alpha, alpha, beta, beta, alpha, beta, beta, alpha.
19 The uranium 235 natural radioactive series ends with a stable isotope of lead, lead 207. What is the total number of alpha and beta particles emitted in this series?
20 Verify the comparison in the chapter of the relative sizes of the nucleus and the atom. If the period representing the nucleus has a diameter of 1.0×10^{-8} cm, calculate in yards the diameter of the atom in which it would exist.
21 Calculate the approximate density of the nucleus of the lithium atom if its mass is 1.2×10^{-23} g and the radius of the nucleus is approximately 1.0×10^{-13} cm. (The volume of a sphere = $\frac{4}{3}\pi r^3$.)

APPENDIX A

APPENDIX A EXPONENTIAL NOTATION AND THE STANDARD FORM OF A NUMBER

The numbers a scientist works with are often very large or very small. The distance of 93,000,000 miles from the earth to the sun and the length of 0.00000001 cm for the radius of an atom are awkward numbers to write in ordinary decimal notation and even more awkward to handle in numerical calculations. Such numbers are just as meaningful and much more usable if written in *exponential form*. The exponential form of a number is frequently referred to as the *standard form* of a number. It consists of two factors. The first factor has a value between 1.00 ⋯ and 9.999 ⋯, and the second factor is 10 raised to an appropriate power. In standard form the distance to the sun would be written 9.3×10^7 miles and the radius of the atom 1×10^{-8} cm.

The use of numbers in standard form simplifies calculations. The student may need to review a few rules and definitions for handling exponents, however. The four rules and three definitions given below are illustrated by examples.

RULE 1

$$a^n \times a^m = a^{n+m}$$

To multiply two numbers expressed as powers of the same base, simply add the exponents and apply the result as exponent to the base.

$$2^2 \times 2^3 = 2^{2+3} = 2^5 = 32$$
To check: $4 \times 8 = 32$

RULE 2

$$\frac{a^n}{a^m} = a^{n-m} \qquad a \neq 0$$

To divide two numbers expressed as powers of the same base, simply subtract the exponent of the divisor from the exponent of the dividend and apply the result as exponent to the base:

$$\frac{2^3}{2^2} = 2^{3-2} = 2^1 = 2$$
To check: $\frac{8}{4} = 2$

RULE 3

$$(a \times b)^n = a^n \times b^n$$

The power of a product may be expressed as the product of each factor raised to that power:

$$(2 \times 3)^2 = 2^2 \times 3^2 = 4 \times 9 = 36$$
$$\text{To check: } 6^2 = 36$$

RULE 4

$$(a^n)^m = a^{n \times m}$$

The exponent of the power of a power is the product of the exponents.

$$(2^3)^2 = 2^{3 \times 2} = 2^6 = 64$$
$$\text{To check: } 8^2 = 64$$

DEFINITION 1

$$a^0 = 1 \quad a \neq 0$$

Any number raised to the zero power is equal to 1. This is a consequence of Rule 2. If

$$\frac{a^2}{a^2} = a^{2-2} = a^0$$

and

$$\frac{a^2}{a^2} = 1$$

then

$$a^0 = 1$$

DEFINITION 2

$$a^{-n} = \frac{1}{a^n}$$

This definition also follows upon Rule 2. If

$$\frac{a^2}{a^3} = a^{2-3} = a^{-1}$$

and

$$\frac{a^2}{a^3} = \frac{1}{a}$$

APPENDIX A

then
$$a^{-1} = \frac{1}{a}$$

DEFINITION 3
$$a^{1/n} = \sqrt[n]{a}$$

This definition establishes the notation for indicating the roots of a number.

The following calculations make use of the standard form of numbers involved and apply the above rules and definitions.

Example A-1
$$\frac{300 \times 0.000200}{250}$$

Express in standard form:
$$\frac{(3.00 \times 10^2)(2.00 \times 10^{-4})}{2.50 \times 10^2}$$

Simplify by collecting numerical factors and tens factors:
$$\frac{3.00 \times 2.00}{2.50} \times \frac{10^2 \times 10^{-4}}{10^2} = 2.40 \times 10^{-4}$$

Example A-2
$$\frac{0.120 \times 0.00830}{390}$$

Express in standard form:
$$\frac{(1.20 \times 10^{-1})(8.30 \times 10^{-3})}{3.90 \times 10^2}$$

Simplify by collecting numerical factors and tens factors:
$$\frac{1.20 \times 8.30}{3.90} \times \frac{10^{-1} \times 10^{-3}}{10^2} = 2.55 \times 10^{-6}$$

Example A-3
$$\sqrt{1.44 \times 10^4} = (1.44 \times 10^4)^{1/2} = 1.2 \times 10^2$$

Example A-4
$$\sqrt{\frac{7{,}860}{0.314}} = \left(\frac{7.86 \times 10^3}{3.14 \times 10^{-1}}\right)^{1/2} = \left(\frac{7.86}{3.14}\right)^{1/2} \times \left(\frac{10^3}{10^{-1}}\right)^{1/2}$$
$$= 1.58 \times 10^2$$

APPENDIX B

APPENDIX B   MEASUREMENT AND UNCERTAINTY

Some uncertainty is associated with every physical measurement. The finer the calibration of a measuring instrument, the less the uncertainty per measurement. Consider three instruments for measuring length: (1) a meterstick calibrated only in decimeters, (2) a second meterstick with calibrations in centimeters, and (3) a metric ruler with calibrations in centimeters and millimeters. The length of an object measured with the first stick is found to be about halfway between 2 and 3 dm. This measurement is recorded as 2.5 dm. The second meterstick is used to measure the same object, and it is found that the length is not, in fact, quite 2.5 dm but is considerably more than 2.4 dm, so this measurement is recorded as 2.47 dm. Then the metric ruler is used, and this time the length is recorded as 2.472 dm. Each instrument, with a finer calibration than the one preceding, permits the measurement to be recorded to one more decimal place. Notice, however, that for each measurement, the last digit recorded is an estimated one.

The same sequential order of precision is mentioned on page 7 in the description of the three laboratory balances. These, as described, weigh objects to the nearest 0.1, 0.01, and 0.001 g, respectively.

If, from readings, the range within which the value of the estimated digit must lie is apparent, this range may be given as part of the measurement. The first reading, above, might have been 2.5 $\pm$ 0.2 dm; the second might have been 2.47 $\pm$ 0.01 dm, and so on. We are seldom in a position to state the range this definitely, however, because this requires a large number of readings. In this text we have adopted the method for expressing the uncertainty of a measurement known as *significant digits*. By this method, each digit in a number is taken to be exact except the last, which is an estimated one.

Significant digits must be differentiated from *place-holding zeros* in a number. The rule for determining the number of significant digits in a number is this: count from left to right, starting with the first nonzero digit and counting all digits to the end of the number. A measurement of 8.24 cm has three significant digits. Whether that number is re-expressed as 0.824 cm or 0.00824 m, it still has the same number of significant digits. The location of the decimal place in a number does not affect the number of significant digits. Some examples follow.

Number	Number of Significant Digits
2,345	4
102	3
0.00345	3
2.3	2
0.0001	1

Numbers like 2,300 are ambiguous. It is possible, for instance, if a bank balance is meant, that this is a four-significant-digit number, meaning $2,300 and some cents. On the other hand, it may be a very rough measurement of distance and have only two significant digits. Here is another advantage to the use of the standard form for numbers in scientific work. There can be no ambiguity about a number in standard form. If the number has two significant digits, it is written 2.3×10^3. If it has four significant digits, it is written 2.300×10^3.

"Perfect numbers" have no uncertainty involved in their expression at all. They may be perfect because so defined, as 12 in./foot, or 60 sec/min, or they may be counting numbers, such as 3 samples or 8 test tubes.

Using one inaccurate measurement in a calculation obviously gives a calculated result no more accurate than that least accurate item. A simple example should make this clear enough. Suppose the side of a perfect square is measured as 2.3 ft. The area found by multiplying 2.3 ft × 2.3 ft is 5.29 ft². Expressed this way, this number implies that the area is no smaller than 5.285 ft² and no larger than 5.294 ft². But the linear measurement could have been anything between 2.25 and 2.34 ft. Squaring each of these lengths gives areas of 5.06 and 5.48 ft², respectively. Both are possible, according to the uncertainty involved in the linear measurement. Thus, instead of knowing the area in two exact digits and one estimated, as the answer 5.29 ft² implies, we actually know it only in one exact digit and one estimated digit, and it should be written 5.3 ft².

Three rules are all that are necessary for the application of the method of significant digits to calculations.

1. For multiplication and division, express the result in the same number of significant digits as the item with the fewest significant digits.
2. For addition or subtraction, write digits in decimal places in the result as far to the right as you find that all items being added or subtracted have digits.
3. Before beginning a calculation, determine the number of significant digits in which the answer will be expressed and then round off each value to *one extra* significant digit before calculation. The result, then, must be rounded off again to the correct number of digits.

Some examples follow.

Example B-1 Determine the density of a cylindrical object if its dimensions are length = 4.34 cm, diameter of base = 2.2 cm, mass = 57.329 g. We see that the result is limited to two significant digits by the recorded value for the

APPENDIX B

diameter. Therefore the other two values are rounded off to three significant digits before calculating.

$$\text{Density} = \frac{57.3 \text{ g}}{(2.2/2)^2(3.14)(4.34) \text{ cm}^3} = 3.47 \text{ g/cm}^3$$

And then this result is expressed as 3.5 g/cm^3. *Note:* The three-significant-digit value for π is chosen by the same criterion.

Example B-2 Add the following linear measurements and express the result in centimeters

$$2.34 \times 10^{-2} \text{ mm} \qquad 1.58 \times 10^{-2} \text{ m} \qquad 7.62 \times 10^{-1} \text{ cm}$$

$$\begin{aligned}
2.34 \times 10^{-2} \text{ mm} &= 2.34 \times 10^{-3} \text{ cm} = 0.00234 \text{ cm} \\
1.58 \times 10^{-2} \text{ m} &= 1.58 \text{ cm} \quad\quad\quad\, = 1.58 \text{ cm} \\
7.62 \times 10^{-1} \text{ cm} &= 7.62 \times 10^{-1} \text{ cm} = \underline{0.762 \text{ cm}} \\
&\quad\quad\quad\quad\quad\quad\quad\quad\quad\quad\quad\; 2.34 \text{ cm}
\end{aligned}$$

Result: 2.34 cm.

APPENDIX C

APPENDIX C UNITS OF MEASUREMENT

Mass

Metric	English
Kilogram = 1×10^3 g	Ton = 2×10^3 lb
Hectogram = 1×10^2 g	*Pound*
Decagram = 1×10 g	Ounce = $\frac{1}{16}$ lb
Gram	
Decigram = 1×10^{-1} g	
Centigram = 1×10^{-2} g	
Milligram = 1×10^{-3} g	
Microgram = 1×10^{-6} g	

Conversion Factors

1 lb = 454 g
1 oz = 28.4 g
1 kg = 2.21 lb

Length

Metric[1]	English
Kilometer = 1×10^3 m	Mile = 5.28×10^3 ft
Hectometer = 1×10^2 m	Yard = 3 ft
Decameter = 1×10 m	Foot = 12 in.
Meter	*Inch*
Decimeter = 1×10^{-1} m	
Centimeter = 1×10^{-2} m	
Millimeter = 1×10^{-3} m	
Micrometer = 1×10^{-6} m = 1×10^{-3} mm	
Nanometer = 1×10^{-9} m = 1×10^{-6} mm	
Angstrom = 1×10^{-10} m = 1×10^{-7} mm	

Conversion Factors

1 in. = 2.54 cm
1 m = 39.4 cm
1 km = 0.621 mile

Volume

Metric	English
Kiloliter = 1×10^3 liters	Gallon = 4 qt
Hectoliter = 1×10^2 liters	*Quart*
Decaliter = 1×10 liters	Pint = $\frac{1}{2}$ qt

[1] The old terms micron, for micrometer, and millimicron, for nanometer, appear in some texts.

Liter

Deciliter $= 1 \times 10^{-1}$ liter
Centiliter $= 1 \times 10^{-2}$ liter
Milliliter $= 1 \times 10^{-3}$ liter

Conversion Factors

1 qt = 0.946 liter
1 liter = 1.06 qt
Definition 1 liter = 1 dm³
therefore 1 ml = 1 cm³

Heat Intensity

	Fahrenheit	Celsius
Freezing point of water:	32°F	0.0°C
Boiling point of water:	212°F	100°C

Conversion Formulas

$F° = \frac{9}{5}C° + 32$
$C° = \frac{5}{9}(F° - 32)$

APPENDIX D

APPENDIX D   CHEMICAL NOMENCLATURE

Over 2 million compounds have been identified and characterized. Each has been assigned a name. Thus chemical nomenclature must obviously be highly organized and systematic. This discussion is not sufficient to enable you to name all 2 million compounds, but when you have mastered it, you will be able to name most of the compounds discussed in this text.

Ions and Ionic Compounds

The name of the ions of a metal is for the most part the name of the metal itself, except when the metal forms more than one ion. Then it is necessary to indicate the charge on the ion by placing a Roman numeral in parenthesis after the name of the metal. Thus we have:

Sodium ion for Na^+
Calcium ion for Ca^{++}
Aluminum ion for Al^{3+}
Iron(II) ion for Fe^{++}
Iron(III) ion for Fe^{3+}
Chromium(II) ion for Cr^{++}
Chromium(III) ion for Cr^{3+}

The reader will recall that formation of ions with more than one ionic charge is primarily a property of the transition metals. One of the few common positive ions not containing a metal is the NH_4^+ ion. It is named by dropping the final *-a* from ammonia and adding *-um*.

Negative ions containing only one element are named by combining the suffix *-ide* with the stem of the name of the element. The stems derived from the names of some common elements are listed below:

Boron	bor-
Bromine	brom-
Carbon	carb-
Chlorine	chlor-
Fluorine	fluor-
Iodine	iod-
Nitrogen	nitr-
Oxygen	ox-
Sulfur	sulf-

Thus we have

Chloride ion for Cl^-
Sulfide ion for S^{--}
Nitride ion for N^{3-}

When oxygen is present in the negative ion, the relative number of oxygen atoms is indicated with prefixes and suffixes. The common prefix and suffix combinations are listed below in order of decreasing oxygen content of the ion.

> Per . . . . ate
> . . . . ate
> . . . . ite
> Hypo . . . . ite

Thus we have

> Perchlorate for ClO_4^-
> Chlorate for ClO_3^-
> Chlorite for ClO_2^-
> Hypochlorite for ClO^-

Note that these prefix and suffix combinations indicate a *relative* number of oxygen atoms and not a definite number of oxygen atoms. For example, as names for the sulfur-oxygen ions we have

> Hyposulfite ion for SO_2^{--}
> Sulfite ion for SO_3^{--}
> Sulfate ion for SO_4^{--}
> Persulfate ion for SO_5^{--}

In naming ionic compounds, one simply combines the names of the two ions, naming the positive ion first:

> Ammonium chloride for NH_4Cl
> Potassium sulfate for K_2SO_4
> Iron(II) bromide for $FeBr_2$
> Iron(III) bromide for $FeBr_3$

Compounds of Nonmetals   The number of atoms of each element is designated by the appropriate Greek prefix preceding the name of the constituent. Commonly used Greek prefixes are

> mono-   hexa-
> di-   hepta-
> tri-   octa-
> tetra-   ennea-
> penta-   deca-

The prefix *mono-* is not always included. That is, if no prefix is designated, *mono-* is understood. The element which appears first in the formula is named first. The second element in the formula is designated by combining the stem of the name of the element with the suffix *-ide:*

CO	carbon monoxide
CO_2	carbon dioxide
CCl_4	carbon tetrachloride
SO_2	sulfur dioxide
SO_3	sulfur trioxide
N_2O_3	dinitrogen trioxide
P_4O_6	tetraphosphorus hexaoxide

Aqueous Acids

For aqueous acids formed from binary hydrogen compounds, the name is determined by attaching the prefix *hydro-* and the suffix *-ic* to the appropriate stem (see page 261). This term is followed by the word *acid*. Thus we have hydrochloric acid for hydrogen chloride, HCl, dissolved in water, hydrobromic acid for hydrogen bromide dissolved in water, and hydrofluoric acid for hydrogen fluoride dissolved in water.

In determining the names for ternary aqueous acids, the prefix *hydro-* is omitted. The suffixes *-ous* and *-ic* are substituted for the *-ite* and *-ate* endings of the names of the analogous ions, and the word *acid* is added. Thus we have perchloric acid for $HClO_4$, sulfuric acid for H_2SO_4, and sulfurous acid for H_2SO_3.

INDEX

Absolute temperature scale, 127
Absolute zero, 127
Accelerators, 234
Acetic acid, 104, 187, 192
Acetylene, 100
Acids:
 aqueous, 182-183
 conjugate, 188-189
 ionic, 187
 organic, 104
 polyprotic, 186-187
 strength of, 188-189
 strong, 181
 weak, 181
Activated complex, 215-221
Activational energy, 215, 221
Air pollution, 192, 205-206
Alcohol, 104
Alkali metals, 54
Alkaline earth metals, 54
Allotropy, 94
Alpha particles, 29, 30, 231
 scattering of, 29
Ammonia, 192
 hydrogen bonding in, 139
 manufacture of, 222
 molecule of, 73-76
Ammonium chloride, 203
 electron formula of, 79
Ammonium ion, 187
 electron formula of, 79
Ammonium polysulfide, 93
Amorphous carbon, 97
"Amorphous" solids, 154
Ampere, 20
Amphiboles, 111, 112
Aniline, 103
Anode, 201
Antacid, 191
Antimony, 96
Aqueous acids, 182-183
Aqueous bases, 182-184
Argon, 208
Aristotle, 19
Aromatic compounds, 103
Arsenic, 96
Atmospheric pressure, 123
Atomic energy, 239
Atomic number, 31
Atomic theory, 29
Atomic weight, 31-32

Atoms, 29-42
 nuclei of, 29, 31
 radioactive, 29
 relative sizes, 56
Avogadro, Amadeo, 132
Avogadro's hypothesis, 132
Avogadro's number, 133

Baking powder, 191
Baking soda, 191
Balance:
 platform, 7
 top-loading, 8
 triple-beam, 8
Barometer, 123-124
Bases:
 aqueous, 182-184
 conjugate, 188-189
 strength of, 188
 structure of, 181
 weak, 182
Benzene, 102-103
Benzene ring, 101
Benzpyrene, 104
Beryl, 111, 112
Beryllium chloride, 80-81
 electron formula of, 80-81
Beta particle, 29, 231
Bicarbonate ion, 190-191
Binding energy, 238
Blood, 191-192
Bohr, Niels, 34
Boiling, 128-129, 136
Boiling point, 12, 128
Bond, 49
 coordinate covalent, 78
 covalent, 49, 69-73
 double, 82, 99, 101
 electrovalent, 49
 ionic, 49, 69
 multiple covalent, 82, 99
 nonpolar covalent, 71
 percent ionic character, 73
 pi, 82, 101, 158
 polar covalent, 70-74
 rotation of, 99
 sigma, 82, 101
 triple, 82, 99
Bond angles, 73-77
Bond lengths, 82

265

INDEX

Bond orbital, 69
Bond strengths, 82
Borates, 114–115
Boron, 114
Boron trifluoride, electron formula of, 80
Boyle, Robert, 124
Boyle's Law, 124
Bromine pentafluoride, electron formula of, 81
Buffers, 191
Butane, 105

Calorie, 13
  large, 13
  small, 13
Canal rays, 24
Cancer, relation to smoking, 103–104
Carbon, 93
  allotropes, 96
  amorphous forms, 97, 154
  atomic weight standard, 31–32
Carbon dioxide, 148, 190–192
  electron formula of, 82
Carbon monoxide, electron formula of, 82, 84
Carbonate ion, 83, 190
Carbonic acid, 187, 190–192
Carcinogens, 103
Catalyst poisons, 224
Catalysts, 221
  applications of, 222–224
  contact, 223
Catalytic muffler, 224
Cathode, 21, 201
Cathode ray tube, 21, 34
Cathode rays, 21, 25
Cellulose, 104
Celsius scale, 12
Chadwick, James, 234
Chain reaction, 218
Chalcogens, 54
Charles, Jacques, 128
Charles' law, 128
Chemical activity and periodic table, 58
Chemical bond, 49, 69–73
Chemical change, 14, 163
Chemical energy, 164
Chemical equations, 173–174
Chemical reactions:
  combination, 164
  endothermic, 164
  exothermic, 164

Chlorine, 109
Chlorine dioxide, 218
Closest-packed structure, 156
Coal tar, 104
Compounds, 5
Condensation, 134
Conductors, 157
Conjugate acids, 188–189
Conjugate bases, 188–189
Contact catalyst, 223
Coordinate covalent bond, 78
Copper, 157, 207
Coulomb, 20
Covalent bond, 49, 69–73, 215
Critical mass, 237
Critical temperature, 134
Cryochemistry, 218
Crystal energy, 168
Crystal lattice, 154–157
  body-centered, 155
  face-centered, 155
  hexagonal, 155
  simple cubic, 155
Crystals:
  cleavage of, 156
  closest-packed, 156
Cyclohexane, 98, 99, 102
Cyclopentane, 98, 99

Dalton, John, 19
  and atomic theory, 29, 31
  and gases, 124
  and symbols, 32
Dalton's law of partial pressures, 125–126
Democritus, 19
Density, 7, 10, 121, 122, 147
  comparison of three physical states, 121, 122
  of gases, 11
Detergents, 176
Deuteron, 239
Diamond, 96–97, 153–154, 157
Dichlorine heptaoxide, 109
Dichloroethane, isomers of, 106
Dichloroethene, isomers of, 106–107
Dimethyl ether, 104
Dipole, 71
Dissociation, 182
Double bond, 82, 99, 101
Dry cell, 203–204
Dry Ice, 148

INDEX

Edison, Thomas A., 22
Elastic collision, 130
Electric conductors, 157
Electric current, 20, 23
Electric energy, 130
Electricity, static, 19-20
Electrochemistry, 200-203
Electrodes, 201
Electrolysis, 202
Electrolyte, 202
Electrolytic cell, 202
Electron, 21-23, 25, 31
   charge of, 22, 25
   mass of, 22, 25
Electron affinity, 56, 71
Electron configurations, 37-43
   stable, 58
Electron exchange reactions, 199
Electron formulas, 54-55, 69, 85
Electron spin, 36
Electron theory of the atom, 33-43
Electronegativity, 71-73
Electrons:
   unpaired, 84
   unshared pair, 76
Electroplating, 206, 207
Electrostatic forces, 230
Electrovalent bond, 49
Element 104, 235
Element 105, 235
Elements, 5, 49
   man-made, 234-237
Empirical formulas, 56
Endothermic reactions, 164
Energy, 4
   chemical, 164
   electric, 130
   heat, 12
   kinetic, 130
   potential, 130, 164-166
Energy level, 34, 35
Entropy, 164-166
Equations, chemical, 173-174
Equilibrium, 135
   acid-base, 184
   shifting of, 185-186
Ethane, 98-100
Ethene, 100
   molecule of, 101
Ether, 104
Ethyl alcohol, 104
Ethylene, 100
Ethylene dibromide, 205

Ethyne, 100
Evaporation, 128-129, 134
Exhaust gases, 205
Exothermic reactions, 164
Exponential form of a number, 245
Exponents, rules for, 245-247

Fahrenheit scale, 12
Feldspars, 113
Fertilizers and lake pollution, 213-214, 223
Filling order, subenergy levels, 37, 38
Fireworks, 34
Fission reactions, 237-240
Fixed nitrogen, 222
Fluids, 121-140
Fluorine, standard for electronegativity, 71
Fluorochlorobromomethane, 107
Food preservation, 223
Force, 123
Formulas:
   electron, 54
   empirical, 56
   molecular, 70
   simplest, 56
   structural, 98
Franklin, Benjamin, 20
Free energy, 166-167
Free radical reactions, 218
Free radicals, 218
   in smog, 206
Freezing, 134
Freezing point, 129
Fuel cells, 205-206
Fundamental particles, 19-25
Fusion reactions, 237-240

Gamma rays, 232
Gas laws, 123-128
   deviations from, 128
Gases, 11, 123
   diffusion of, 121
   ease of compression, 121
   standard conditions, 124
   velocity of molecules, 131
Gasoline, antiknock properties, 223
Geiger counter, 231
Geometric isomerism, 106, 107
Glucose, 105

Gold, 206
   and Rutherford experiment, 30
Graduated cylinder, 9
Gram, 6-7
   definition of, 151
Gram atomic weight, 32
Gram molecular volume, 133
Gram molecular weight, 133
Graphite, 96-98, 157-158
Gravitational forces, 229

Haber, Fritz, 222
Haber process, 222-223
Half-life, 232
Halogens, 54, 93
Hard water, 175-176
Heat, 12
   of fusion, 150
   intensity of, 12
   quantity of, 12
   of vaporization, 137
Heat effects, 11
Heterogeneous materials, 4
Heterogeneous reactions, 208-209
Heteronuclear bonding, 108
Hexane, 102
Homogeneous materials, 4
Homonuclear bond, 93
Hund's principle of maximum multiplicity, 37
Hybrid orbitals, 75-76
Hydration, 168
Hydration energy, 168
Hydrazine, 206
Hydrocarbons, 104
   constituents of smog, 205
Hydrogen, 93, 156, 215
   in manufacture of ammonia, 222
Hydrogen bomb, 239
Hydrogen bond, 139-140, 149-151
Hydrogen chloride, formation of, 216, 218
Hydrogen fluoride, hydrogen bonding in, 139
Hydrogen iodide, formation of, 215-217
Hydrogen ion, 184
Hydrogen-oxygen fuel cells, 206
Hydronium ion, 184
Hydroxide ion, 182

Ice, hydrogen bonding in, 149
Inhibitor, 224

Inner transition elements, 53
Insulators, 157
Intermolecular attractions, 137-140
Iodine, 215
Ionic acids, 187
Ionic bond, 49, 54-60, 69
Ionic crystals, 54, 152-153
Ionic solids, 152-153
Ionization, 182
Ionization energy, 56, 57, 71
Ionization potential, 56
Ions, 49-60
   and periodic table, 59
   polyatomic, 69-85
Isobutane, 105
Isoelectronic structures, 84
Isomerism, 105
   optical, 108
   stereo-, 105
   structural, 105
Isotopes, 31, 230

Kelvin, Lord, 127
Kelvin scale, 127
Kinetic energy, 130
Kinetic molecular theory, 129-132, 148

"Last" electron, 51
Lead, 204-205
Lead (IV) oxide, 204-205
Lead storage cell, 204-205
Leucippus, 19
Liquids, 11, 128-129
   supercooled, 154
   surface tension of, 129
Liter, 6-7

Magnesium, 207-208
Manganese (IV) oxide, 203
Mass, 6
Mass number, 31
Matter:
   continuity versus discontinuity, 19
   properties of, 4
Measurement, 6, 251
   units of, 257-258
Mechanism of a reaction, 215
Melting, 12
Melting point, 12
Metallic bond, 153
Metallic solids, 153

INDEX

Metalloids, 54
Metallurgy, 206-208
Metals, 53, 153
   earliest known, 3
   electron configuration of, 53
   as reducing agents, 199-200
Metaphosphoric acid, 93, 110, 111
Meter, 6-7
Methane, 98
   molecule of, 73-76
Methyl alcohol, 104
Methyl ethyl ether, 104
Metric system, 6-7
Micas, 113
Mixtures:
   heterogeneous, 4
   homogeneous, 4
   separation of, 5
Molarity, 172
Mole, 132
Molecular formulas, 70
Molecular geometry, 74, 77
   bent, 74, 77
   linear, 74, 77
   tetrahedral, 74, 77
   triangular, 74, 77
   triangular pyramidal, 74, 77
Molecular solids, 148-152
Molecular weight, 132-133
Molecules, 69
   discrete, 69-85
   polar, 74
   polyatomic, 73
Multiple covalent bond, 82, 100

$n + \ell$ value, 37, 38
Naming of chemical compounds, 261-263
Naphthalene, 103
Neon lights, 34
Neptunium, 235
Network solids, 153-154
Neutralization, 190
Neutron, 24-25, 31, 229
Nitrite ion, electron formula of, 83
Nitrogen, 93
   in the manufacture of ammonia, 222
Nitrogen dioxide, 218
   as a catalyst, 222
   electron formula of, 84
   in smog, 206
Nitrogen oxide, 218
Noble metals, 199

Nomenclature, 261-263
Nonmetals, 53
   electron configuration of, 53
Nonpolar covalent bond, 71
Nuclear atom, 29
Nuclear equations, 230
Nuclear forces, 230
Nuclear power plants, 239
Nuclear reactions, 230
Nuclear symbols, 230
Nuclear transformations, artificial, 232-240
Nucleon, 31, 229
Nucleus of atom, 31, 229

Octet rule, 71, 79
   expansion of the octet, 81
Odd molecules, 218
Optical isomerism, 106-108
Orbitals, 35-36
   atomic versus molecular, 70
   hybrid, 75
   shapes of, 36
Ores, 207
Organic chemistry, 78, 97
Orthosilicate ion, 111
Oxalic acid, 187
Oxidation-reduction reactions, 199-200
Oxidizing agent, 199-200
Oxygen, 93
   and homonuclear bonding, 95
   isotopes of, 31
   in manufacture of sulfuric acid, 222
   electron formula of, 84
Ozone in smog, 206

Paramagnetism, 84
Partial pressure, 126
Pauli, W., 37
   and exclusion principle, 37
Perfect numbers, 252
Periodic classification, 49-54
Periodic law, 49
Periodic table, 49, 50, 52, 57-59
   and chemical activity, 58
   families, 54
   groups, 51
   periods, 51
Periodic trends, 58
Petroleum, 5
Phenol, 103
Phosphates, 176
Phosphorescence, 231

Phosphoric acid, 187
Phosphors, 29-30
Phosphorus, 93, 110
  allotropes, 95-96
Phosphorus pentafluoride, electron formula of, 81
Photoelectric cell, 23
Physical change, 14, 163
Physical equilibrium, 135
Physical states, 4, 10-11
  gas, 11
  liquid, 11
  solid, 11
Pi bond, 82, 101, 158
Platform balance, 7
Plutonium, 235
Polar covalent bond, 70-74
Polar molecule, 74
Polarity, 138
Pollution:
  air, 192
  and radioactivity, 239-240
Polonium, 29
Polymer, 93
Polymeric ions, 93
Polymeric molecules, 93
Polyprotic acids, 186-187
Polysulfide ions, 95
Polywater, 151-152
Positive rays, 24, 25
Positron, 234
Potassium chlorate, electron formula of, 78
Potassium chlorite, electron formula of, 79
Potassium hypochlorite, electron formula of, 79
Potassium perchlorate, electron formula of, 78
Potential energy, 130, 164-166
Precision, 251
Pressure, 123-124
Propane, 98, 99
Propanoic acid, 104
Properties, 3, 4
  chemical, 6, 14
  physical, 6, 14
Proton, 24, 25, 31, 181, 229
  charge of, 24, 25
  mass of, 24, 25
Psuedo-solids, 154
Pure substances, 5
  compounds, 5

Pure substances:
  elements, 5
Pyrosilicate ion, 111
Pyrosulfate ion, 109
Pyroxenes, 111, 112

Quantum mechanics, 34
Quantum numbers, 35-39
Quartz, 113, 114

Radioactive atoms, 29
Radioactive wastes, disposal of, 239
Radioactivity:
  artificial, 233-234
  natural, 231
Rare earth elements, 53
Rare gases, 53, 54, 93
  electron configuration of, 53
Rate of reaction, 208
  factors affecting, 208-210, 219-221
Reactions:
  chain, 218
  electron exchange, 199
  free radical, 218
  heterogeneous, 208-209
  mechanism of, 215
  oxidation-reduction, 199-200
Reducing agent, 199-200
Refining, 207
Representative elements, 53
Resonance, 82
Resonance hybrid, 83
Rider, balance, 6
Rule of eight, 71, 79
Rutherford, Ernest, 29, 229, 232

Saturated solution, 171
Selenium, 93, 95
Sigma bond, 82, 101
Significant digits, 251-253
Silicates, 110-113, 176
Silicon, 93, 108, 110
Silicon dioxide, 113-114
Silver, 206
Simplest formulas, 56
Smog, 192, 205-206, 218
Soap, 174-176
Sodium carbonate, electron formula of, 83
Sodium chloride, 156
Sodium nitrite, electron formula of, 83

INDEX

Sodium sulfate, electron formula of, 77
Sodium sulfite, electron formula of, 78
Solids, 11
   classes of, 147
Solubility, 170–171
Solute, 171
Solutions, 5
   concentration of, 172
   dissolving process, 167–170
   ions in, 173
   saturated, 171
   supersaturated, 171
   terminology of, 170–172
   unsaturated, 171
Solvent, 171
sp-orbitals, 80
sp^2-orbitals, 80, 101
sp^3-orbitals, 76–78
sp^3d-orbitals, 81
sp^3d^2-orbitals, 81
sp^3d^3-orbitals, 81
Specific gravity, 10
Specific heat, 14, 151
Spectroscopy, 49
Standard conditions for gases, 124
Standard form of a number, 245
Stereoisomerism, 105
   geometric, 106
   optical, 106, 107
Stoichiometry, 225–226
Storage cell, 204–205
Strong acids, 181
Strontium 90, 239
Structural formulas, 98
Structural isomerism, 105
Subenergy level, 35
Sublimation, 148
Sugar, 105
Sulfate ion, electron formula of, 77–78
Sulfide ion, 93
Sulfite ion, electron formula of, 78
Sulfur, 93, 109–110
   allotropes, 93–95
   monoclinic, 93
   orthorhombic, 93
Sulfur dioxide, 192
   constituent of smog, 205
   electron formula of, 82–83
   in the manufacture of sulfuric acid, 222
Sulfur hexafluoride, electron formula of, 81
Sulfur trioxide, 109–110, 192

Sulfur trioxide:
   electron formula of, 83
Sulfuric acid, 192
   in lead storage cell, 204–205
   manufacture of, 222
Sulfurous acid, 187
Supercooled liquids, 154
Supersaturated solution, 171
Surface tension, 129, 137
Symbols, 32–33

Talc, 113
Television, picture tube, 23
Tellurium, 95
Temperature, 12
   Celsius scale, 12
   Fahrenheit scale, 12
   Kelvin scale, 127
Ternary compounds, 77
Tetraethyllead, 205, 223
Theory, 3
Thermometer, 12
Thomson, J. J., 22
Titanium, 208
Toluene, 103
Top-loading balance, 8
Transition elements, 53, 58
Transuranic elements, 234
Trimethylenemethane, 219
Triple-beam balance, 8
Triple bond, 82, 99
Triton, 239

Uncertainty, 251
Units:
   of measurement, 257–258
   standard, 6–7
Unsaturated solution, 171
Uranium 235, 237

Valence electrons, 54
van der Waals forces, 138, 148
Vapor pressure, 135–137
Vernier caliper, 9
Vinegar, 192
Viscosity, 11
Volt, 20
Voltage, 202
Voltaic cell, 200–202
Volumes of liquids, determination of, 9
Volumes of solids, determination of, 9

Water:
    as an acid or base, 182
    hard, 175-176
    hydrogen bonding in, 139, 149-151, 163
    maximum density temperature of, 150
    molecule of, 73-74, 77
    softening of, 175-176
    specific heat of, 151
Weak acids, 181
Weak bases, 182
Weight, 6
White dwarfs, 229
Wilson cloud chamber, 233

Xenon tetrafluoride, electron formula of, 82

Zeolites, 176
Zinc chloride, 203

INTERNATIONAL ATOMIC WEIGHTS 1969
Based on the assigned relative atomic mass of $^{12}C = 12$

	SYMBOL	ATOMIC NUMBER	ATOMIC WEIGHT
actinium	Ac	89	227
aluminum	Al	13	26.9815
americium	Am	95	243
antimony	Sb	51	121.75
argon	Ar	18	39.948
arsenic	As	33	74.9216
astatine	At	85	210
barium	Ba	56	137.34
berkelium	Bk	97	247
beryllium	Be	4	9.01218
bismuth	Bi	83	208.9806
boron	B	5	10.81
bromine	Br	35	79.904
cadmium	Cd	48	112.40
calcium	Ca	20	40.08
californium	Cf	98	249
carbon	C	6	12.011
cerium	Ce	58	140.12
cesium	Cs	55	132.9055
chlorine	Cl	17	35.453
chromium	Cr	24	51.996
cobalt	Co	27	58.9332
copper	Cu	29	63.546
curium	Cm	96	245
dysprosium	Dy	66	162.50
einsteinium	Es	99	249
erbium	Er	68	167.26
europium	Eu	63	151.96
fermium	Fm	100	255
fluorine	F	9	18.9984
francium	Fr	87	223
gadolinium	Gd	64	157.25
gallium	Ga	31	69.72
germanium	Ge	32	72.59
gold	Au	79	196.9665
hafnium	Hf	72	178.49

Note: The elements in color type are those found in the human body.